Structure and Evolution

of Vertebrates

A Laboratory Text for Comparative Vertebrate Anatomy

ALAN FEDUCCIA

UNIVERSITY OF NORTH CAROLINA

W·W·NORTON & COMPANY · NEW YORK LONDON

To my dad
and Aunt Ang

W. W. Norton & Company, Inc., 500 Fifth Avenue, New York, NY 10110
W. W. Norton & Company Ltd, 10 Coptic Street, London WC1A 1PU

Library of Congress Cataloging in Publication Data

Feduccia, J   Alan.
Structure and evolution of vertebrates.

Bibliography: p.
1. Vertebrates—Anatomy. 2. Anatomy, Comparative.
I. Title. [DNLM: 1. Anatomy, Comparative—Laboratory
manuals. 2. Vertebrates—Anatomy and histology—
Laboratory manuals. QL805 F294L]
QL805.F4 1974   596'.04   74–13968
ISBN 0-393-09291-7

This book was designed by Robert Freese.
The typefaces used are Caledonia and Optima,
set by Fuller Typesetting of Lancaster.
Printed in the United States of America

7   8   9   0

Contents

Preface

Comparative anatomy is a subject that goes back well into the last century. New and interesting details concerning vertebrate evolution emerge each year at a rapid rate, but few of these completely revolutionize our thinking about the major features of vertebrate history. In this sense comparative anatomy is a classical field in biology, yet it is usually a very popular course among the biology undergraduates at most universities. There has always been a great demand for courses in comparative anatomy, not only from biology students, but especially from the many students entering the various allied health sciences where a knowledge of the evolutionary history of the human body is an honored possession.

Many biology departments, in attempts to keep up with modern trends, particularly in areas of molecular biology, have removed comparative anatomy from the core curricula, or in extreme cases have abandoned it completely. This has had unfortunate consequences for students pursuing vertebrate evolutionary morphology or functional morphology, and for students entering the health sciences. In addition, most students who intend to pursue such fields as vertebrate physiology, developmental biology, behavior, psychology, and neurobiology, certainly find it advantageous to have a sound background in comparative anatomy. However, in most curricula time no longer permits the luxury of the lengthy, classical six hours of laboratory a week that has characterized most courses in comparative anatomy. And students may appropriately question the utility of having to learn minute anatomical details, including the "hundred" muscles of the cat.

This text offers an alternative to the classical comparative anatomy laboratory. It attempts to cover the major features of the morphological evolution of the vertebrates, without including the minute detail that has characterized the typical encyclopedic laboratory text.

The courses for which this text is suitable are outlined as follows: First, it may be used as the text for a course taught completely in the laboratory; for example, a one-semester course of one weekly meeting with one hour of lecture preceding three hours of laboratory. Second, it may be used as a laboratory text for comparative anatomy courses that require a three-hour-per-week laboratory; or, with additional material added to each exercise, it may be used for half of the traditional six-hour-per-week laboratory, with detailed dissections used to fill the other session. Third, it may be used to split a six-hour laboratory in which half of the course is developmental biology; many biology departments offer one course combining development and anatomy.

It should also be pointed out that though the animals illustrated are traditional ones for comparative anatomy, one could easily substitute a frog for *Necturus* (mud puppy), and any mammal for the cat, since the great detail in dissection that has characterized most previous texts is replaced by an emphasis on general evolutionary anatomical trends. One of the practical advantages of this type of laboratory course is that the major animals (dogfish, *Necturus*, and cat) are not needed at all until the fifth or sixth week of the course. And by using appropriate demonstrations, laboratory time can be shortened even further.

The first chapter, on vertebrate origins, is intended as an introductory chapter, and because of varying emphasis given in different courses to the protochordates and the lamprey larva, laboratory instructions for the sea squirt, lancelet, and ammocoetes are given at the end of the chapter. Protochordates and the lamprey (Chapter 2) could be covered in one laboratory, or separately, depending on the emphasis desired. Chapter 3, on the evolution of gnathostomes, can be turned into a complete laboratory (or museum trip if possible), incorporated into the lamprey laboratory, or used simply as background information for the student. Chapter 4,

on the cranial skeleton, will generally require two laboratory periods, but the amount of time will vary from course to course depending on the amount of comparative material available and the detail required on the disarticulated skulls. Chapter 5, on the postcranial skeleton, can be covered in one laboratory, or expanded with additional comparative material. Chapters 6 and 7, on the musculature and the nervous system, are perhaps most appropriately covered by splitting three laboratory sessions between each chapter. Chapter 8, on the coelom and pharynx, requires one laboratory; Chapter 9, two periods (if covered completely); and Chapter 10, one period.

I am greatly indebted to Mrs. Yvonne Lee, who skillfully rendered the illustrations. Dr. Douglas M. Lay kindly provided the exercise on a filter-feeding vertebrate for the chapter on vertebrate origins. The excellent micrographs for Figs. 4-1, 4-2, 6-3, and 7-3, were provided by my colleagues Drs. Catherine Henley and Donald P. Costello. Dr. Thomas W. Easton read and criticized various drafts of the manuscript.

To the Student

This is a course centered on the theme of the evolutionary history of the human body, taught primarily from the laboratory. The normal prerequisite for this course is a one-semester or longer course in biology or zoology. It is also assumed that the student has had at least some training in dissection. Therefore, detailed instructions for dissections are not given, since the student will gain more by having the freedom to decide on his or her own methodology of study and dissection. Dissection terms are given in Appendix I.

The phylogeny or evolutionary history of the mammalian body has been one of very gradual change over vast periods of evolutionary time. In a course of this nature it is necessary to

choose only a few animals—a fish, a primitive tetrapod, and a mammal—to illustrate the major evolutionary changes. But these animals also have specialized adaptations that have permitted them to persist into modern times. Thus, while the animals dissected in the laboratory illustrate many of the stages in mammalian evolution, they are not completely representative of the actual ancestors, but may approximate the ancestral condition in many of their structures. When we speak of modern animals as being primitive, we imply that, though the animals themselves are highly specialized, they nonetheless exhibit features approximating those found in the primitive ancestral condition. The process of evolution is mechanistic, and it is only in retrospect that the evolutionary trends can be ascertained. The image of evolutionary progression or evolutionary trends is a completely *a posteriori* view of the continuity of structure that has persisted through geologic time; it does not imply a teleological theme. Organisms have evolved because of genetically caused variations that were highly adaptive to particular sets of selection pressures at particular times in the geologic past.

The phylogeny of vertebrates has been unraveled by studies of fossils through time (paleontology), and also by studies of comparative anatomy and embryology. Comparative evolutionary morphology centers around the homologous features of different organisms. Homologous structures are those that are similar in different organisms because of common ancestry. Studies of embryos have also been very useful in the reconstruction of vertebrate history because the embryo tends to change conservatively, especially in the early stages of its development or ontogeny. Thus by examining the early stages of the ontogeny of an individual, one may find clues to the ancestral condition. For example, the appearance of pharyngeal slits and a notochord in the human embryo tells of our descent from animals having these structures in the adult stage. It is thus primarily through studies of the comparative evolutionary morphology of fossils, and studies of the embryos and adults of living vertebrates, that the evolutionary history of the human body is now understood.

1

Vertebrate Origins

The wide array of organisms like ourselves that are termed vertebrates comprise the **subphylum Vertebrata,** a subdivision of a broader taxonomic category, the **phylum Chordata.** Chordate organisms possess at some time during their life histories a **notochord** as the primary axial skeletal element, and in addition a **dorsal tubular nerve cord** located immediately above the notochord. Also (though not unique to the Chordata), there are **pharyngeal slits** located laterally within the pharynx. These "gill slits" are primitively utilized for feeding (food-straining), later for feeding and gaseous exchange, and finally, in the transition to land, disappear in the adult and appear only transitorily in the embryo during ontogeny. Likewise, the notochord begins its evolution as the most important aspect of the endoskeleton, later becomes reduced with the advent of the vertebral column, and is eventually obliterated almost completely in the "higher vertebrates" by the bony vertebral centrum. Thus, of the three features that are used to characterize chordate organisms, only one, the dorsal tubular nervous system, is fully demonstrable in the adults of all the vertebrate series. All three features, however, may be encountered at some time during the ontogeny or life history of all chordates. Organisms belonging to the **subphylum Vertebrata** may be simply defined as chordates having a vertebral column; this morphological feature is invariably present in all vertebrates.

The "primitive" chordates, collectively called the protochordates, include the hemichordates or **acorn worms** (**Enteropneusta**) and their poorly known cousins, the pterobranchs (**Pterobranchia**), and the **beardworms** (**Pogonophora**), which are often classified as a separate phylum. But among the protochordates the groups that show vertebrate affinity most clearly are the **tunicates** or **sea squirts** (**Urochordata**), and the **lancelets** or "**amphioxi**" (**Cephalochordata**). The probable relationships of these groups to the first vertebrates (**class Agnatha**) and to the echinoderms are illustrated by the phylogeny in **Fig. 1-1.** The major groups of living echinoderms are illustrated in **Fig. 1-2.**

Hemichordates and Pterobranchs

The **subphylum Hemichordata** comprises the Enteropneusta, or acorn worms, the Pterobranchia (for which there is no vernacular name), and the Pogonophora, or beardworms, which are often placed in a separate phylum.

The Pogonophora (see Fig. 1-1) are a poorly known group of strange organisms, for the most part abyssal, tube-dwelling (tubicolous) forms that live in chitinous, secreted tubes. Their long ciliated tentacles gather and digest food,

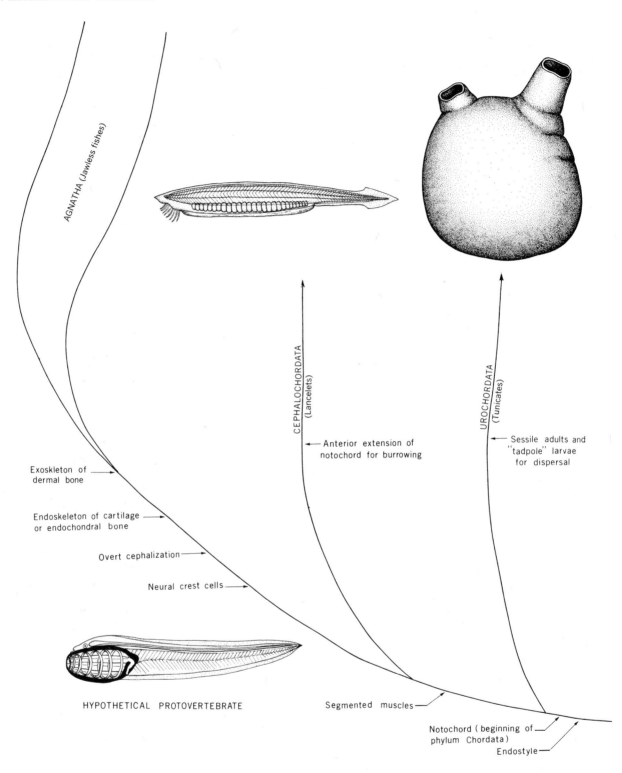

AGNATHA (Jawless fishes)

CEPHALOCHORDATA (Lancelets)

UROCHORDATA (Tunicates)

Exoskleton of dermal bone

Endoskeleton of cartilage or endochondral bone

Overt cephalization

Neural crest cells

Anterior extension of notochord for burrowing

Sessile adults and "tadpole" larvae for dispersal

HYPOTHETICAL PROTOVERTEBRATE

Segmented muscles

Notochord (beginning of phylum Chordata)

Endostyle

Fig. 1-1. Hypothetical phylogeny of the major groups of protochordates. No sequence is intended in this depiction of the four morphological innovations leading to the class Agnatha. (Chart, after Denison; diagrams of enteropneust, pterobranch, and pogonophoran, after Riedl.)

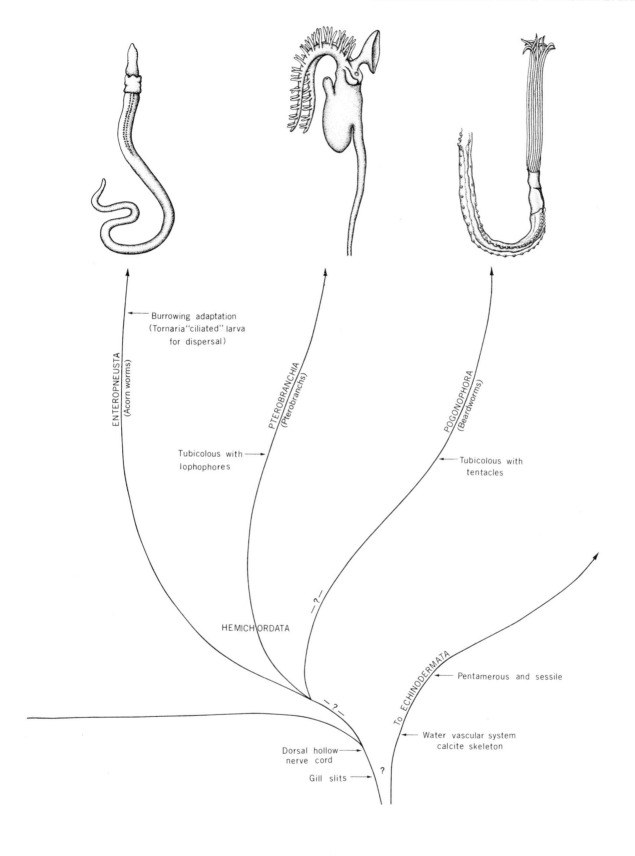

Burrowing adaptation
(Tornaria "ciliated" larva
for dispersal)

ENTEROPNEUSTA
(Acorn worms)

PTEROBRANCHIA
(Pterobranchs)

POGONOPHORA
(Beardworms)

Tubicolous with
lophophores

Tubicolous with
tentacles

HEMICHORDATA

─ ? ─

─ ? ─

To ECHINODERMATA

Pentamerous and sessile

Dorsal hollow
nerve cord

Water vascular system
calcite skeleton

Gill slits

?

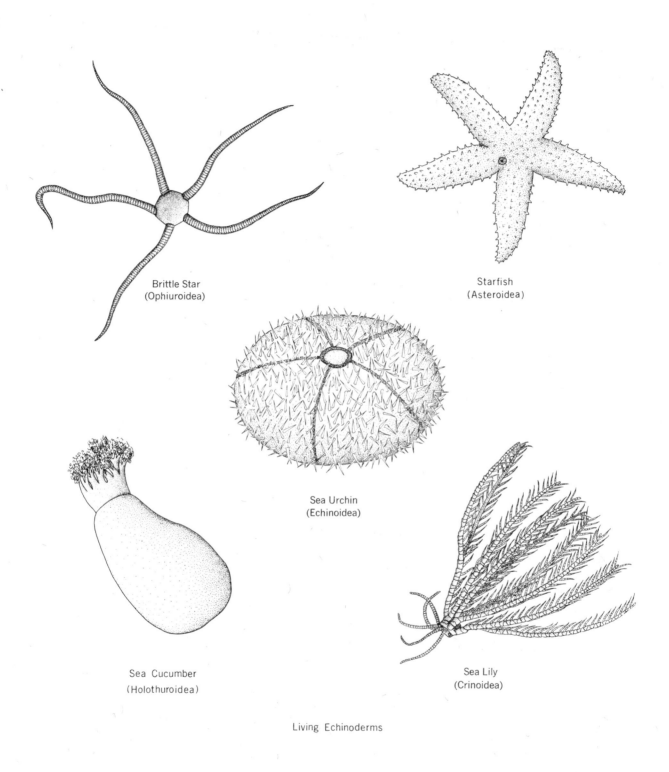

Brittle Star
(Ophiuroidea)

Starfish
(Asteroidea)

Sea Urchin
(Echinoidea)

Sea Cucumber
(Holothuroidea)

Sea Lily
(Crinoidea)

Living Echinoderms

Fig. 1-2. The major groups of living echinoderms.

and there is no trace of a digestive system. The beardworms exhibit so few chordate characteristics that they are frequently regarded as unrelated to the hemichordates. They will not be considered further here.

Pterobranchs (see Fig. 1-1) are very odd creatures, known only from a handful of genera, largely confined to antarctic and subantarctic regions (though they extend into the tropics in the Indo-Pacific region). They are tube-dwelling organisms that form sessile colonies. Individuals project from the tubes and gather food particles from the water by stalked tentacles termed lophophores, which pass the food particles into the mouth via ciliated grooves. Reproduction among the pterobranchs occurs by asexual budding. The pterobranchs appear to be close allies of the acorn worms, their organ systems bearing a very close resemblance, but pterobranchs are so poorly known that it is more profitable to focus attention on the acorn worms.

The acorn worms or enteropneusts (see Fig. 1-1) are the best known of the so-called hemichordate series. Their affinity with the chordates is evidenced by their possession of a dorsal tubular nerve cord and a pharyngeal region with gill slits. They were previously thought to possess a true notochord, but most authors now consider that the "stomochord" is not homologous to the chordate notochord. Most of the acorn worms live in tubes in tidal flats where they gather food in the manner of some "higher" chordates, by filter-feeding. Water containing food particles accumulates in the collar region and is shunted into the mouth by a current created by the strong cilia of the mouth tube. The mouth tube expands into a pharyngeal region (**Fig. 1-3**) where gill slits open to the exterior and expel the water that

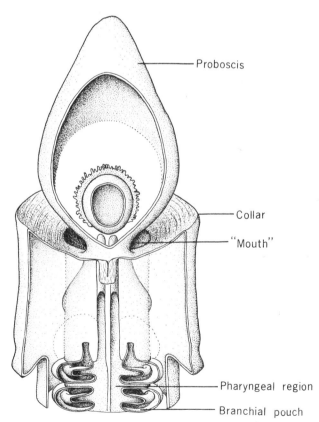

Proboscis

Collar

"Mouth"

Pharyngeal region

Branchial pouch

Fig. 1-3. Modified frontal section of the acorn worm (*Balanoglossus*) to show the morphology of the feeding apparatus. (Modified after Delage and Hérouard.)

has carried food particles into the digestive tube. The digestive tube is a posterior continuation of the pharyngeal tube. This filter-feeding mechanism is essentially that found in the other protochordates, and will be discussed in detail later. Perhaps of greater interest is the general resemblance of the tornaria larva of the acorn worms to the auricularia and bipinnaria larvae of the echinoderms (**Fig. 1-4**).

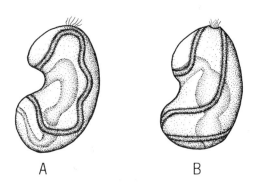

A   B

Fig. 1-4. Larvae of echinoderms (A) and acorn worms (B).

Larval resemblance is additional evidence for the close relationship between the protochordates and the echinoderms. So, though the pogonophorans and pterobranchs are rather poorly known, the affinity of the enteropneusts with the echinoderms seems well established, not only on the basis of larval similarity, but also on the basis of biochemical similarities and the details of coelom formation in both groups.

Garstang was the first student of vertebrate evolution to search for a vertebrate ancestor among the early stages rather than the adults of the invertebrates, and it was he who postulated a theory now called **Garstang's neotenic larva theory** (later elaborated by De-Beer), which has gained wide acceptance among zoologists. The theory rests on the general similarity of the larvae of the hemichordates to those of the echinoderms, along with the general similarity in the ontogeny of certain organ systems. In addition, the theory requires the larvae to linger in their immature state (**neoteny**), and reproduce in the same state, giving rise to sessile adults in the "lower"

series of the protochordates. This theory will be elaborated and continued later in discussions of the Urochordata.

Tunicates

Before developing fully a theory on the origin of vertebrates, attention must be given to a rather curious group of the protochordates. The **tunicates,** or **sea squirts,** of the **subphylum Urochordata** (G., *oura*, tail, indicating the position of the notochord) are a strange group of baglike organisms that, for the most part, occupy themselves by attracting food particles from sea water pulled into the body by an incurrent siphon (Fig. 1-5C). This sea water leaves the organism via an excurrent siphon after the food particles have been trapped; thus the name sea squirt. The other common name, tunicate, refers to the tough "tunic" that surrounds the organism in most adults and is composed of a celluloselike material, tunicin. Tunicates are rather common marine organisms, and a great variety of shapes and sizes are known, ranging from microscopic to several inches; they are found as solitary individuals or as colonies, usually attached to the substrate (Fig. 1-5C). However, there is one group of free-swimming tunicates, the **Larvacea,** which resemble the larvae of other tunicates (Fig. 1-5A), and are generally considered to be neotenic urochordates.

The adults of most of the urochordates show little superficial resemblance to the vertebrates, as there is no real notochord and the only part of the dorsal nervous system remaining is a nerve ganglion that sends out several nerves. Most of the organism (Fig. 1-5C) is composed of a barrel-shaped net, the pharynx, whose job is to filter out food particles from the sea water. Water that has been strained for food particles passes into the atrium. Along the ventral surface of the pharynx is a structure known as the endostyle that secretes a mucoid substance en-

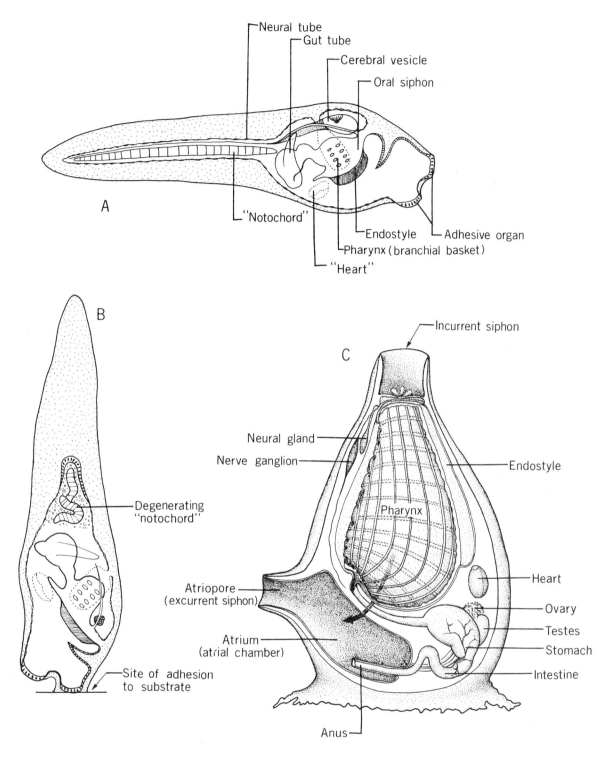

Fig. 1-5. Schematic diagram of the metamorphosis of a solitary tunicate. *A,* the fully developed larva of *Clavelina* shows general chordate body form, with a notochord and a well-developed tail, thus superficially resembling an amphibian tadpole. *B,* the *Clavelina* larva has attached itself to the substrate by use of its adhesive organ, and the "notochord" is seen degenerating as metamorphosis to the sessile adult form occurs. *C,* diagrammatic view of a sessile adult tunicate showing the major morphological features of this visceral, filter-feeding, sessile organism. (*A* and *B,* after Seeliger, in Grassé; *C,* after Delage and Hérouard.)

hancing the entrapment of minute food particles. The resultant food-containing band moves into the esophagus and stomach for digestion. While the sea water passes out the atriopore and excurrent siphon, the collected food particles pass into the intestine and stomach for digestion, and waste products pass out the anus and then out the atriopore with the sea water. Normally tunicates are hermaphrodites, having male and female gonads in the intestinal region, with ducts passing to the atrium. In colonies individuals arise by a budding process, then reproduce sexually. The pharynx is a highly vascularized system; in addition to gathering food, it serves a respiratory function.

Adult tunicates thus reveal little about the possible origins of vertebrates, but in the free-swimming larval stage, a general resemblance to the higher protochordates is clearly evident. The larval forms superficially resemble an amphibian tadpole and are often referred to as **tadpole larvae.** Along the entire length of the tadpole larva's tail is a notochord and a nerve cord immediately dorsal to it (Fig. 1-5A). As in the adult, there is a well-developed pharyngeal region, and water passes through the mouth and gill slits into an atrium, from which water and undigested food particles pass to the exterior along with waste products coming from the intestine. There is a ventral heart that passes blood alternately anteriorly and posteriorly. Upon metamorphosis (Fig. 1-5B), the adhesive organs serve to attach the "tadpole" to the substrate, where it settles down, head first, to become a passive adult, absorbing the tail (and thus the notochord), which is no longer needed as a swimming adaptation. The head of the free-swimming tunicate tadpole thus corresponds to the entire body of the sessile adult.

Where do the tunicates fit into the story of vertebrate evolution? Because the larvae clearly exhibit all the salient chordate characteristics, it is in these forms that we should look for clues to the evolutionary history of the vertebrates. It is probably true that the actual adult ancestor was a sessile, visceral filter-feeding

form, but it is in the larval form that the chordate characteristics appear. There is a tail, which makes possible dispersal of the species, since the "tadpoles" can search for suitable sites to settle down and reproduce. The notochord is no doubt associated with this function. The evolutionary development of higher chordates probably is associated with the retention of the larval habit, and the elimination of the adult stage, again the process of neoteny. The Larvacea (mentioned previously) have done just that. These tunicates have abandoned the adult stage, and persist throughout their entire lives as "mature" larvae, reproducing in that form. They are thus tadpolelike throughout life, and in that sense resemble the group of protochordates we shall next consider.

Cephalochordates

The **cephalochordates** or **lancelets** (see Fig. 1-1) of the **subphylum Cephalochordata** are the protochordates most like vertebrates, and their possession of many "primitive" vertebrate characteristics makes them of special interest to the student of evolution. This subphylum includes a limited variety of species and morphological forms, all referred to commonly as **amphioxus** (G., *amphi*, both, and *oxys*, sharp), an old generic name now unfortunately superseded by ***Branchiostoma*** (there is another genus called ***Asymmetron***). All the species are marine, semitranslucent, fishlike creatures found in coastal waters, and though they possess a fishlike fusiform body, they spend most of their time buried in the sand with only their heads protruding above the substrate. Presumably the anterior extension of the notochord into the head region (hence the name Cephalochordata) is an adaptation for burrowing in the sand. In this position these small animals perform a filter-feeding process very similar to that described for the tunicates, taking in water with food particles, and straining the latter from the medium. From

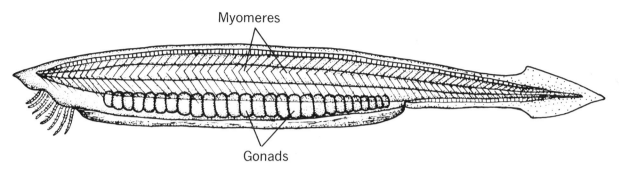

Myomeres

Gonads

Fig. 1-6. Lateral view of an adult lancelet, showing some major external morphological features.

time to time they leave their filter-feeding habitus and actively swim to new sites, presumably in search of new feeding locations, or in attempts to disperse. Because the lancelets are the living protochordates that most closely resemble the probable vertebrate ancestor they deserve our special attention. Amphioxus has often been called the "darling" of the comparative anatomists, because it has most of the characters thought to represent the primitive state among vertebrate animals. The following is a description of the anatomy of a lancelet and an elucidation of its phylogenetic status.

The lancelet body (**Fig. 1-6**) is arranged as a series of muscle blocks, or **myomeres** (G., *myos*, muscle, and *meras*, part), which are covered by a very thin transparent epidermis, and which extend along the trunk; successive myomeres are separated by partitions of connective tis-

sue, the **myosepta** or **myocommata.** The muscle segments are arranged as V-shaped blocks with the apex of the V pointing anteriorly. These are best seen in Fig. 1-6. The segmental arrangement of features that characterizes most vertebrate organisms is known as segmentation, or **metamerism** (G., *meta*, after, and *meras*, part); the segmental arrangement of muscles is termed **myomerism.** Recall that in the other protochordates (including the tunicate tadpole larva) there was little sign of metamerism, while in the lancelets segmentation is one of the obvious features of the body.

Aside from the cartilagelike connective tissue that stiffens the gill region, and the dorsal fin and mouth parts, the only true skeletal element is the **notochord** (G., *noton*, back, and *chorde*, cord), the primitive vertebrate axial skeletal element (**Fig. 1-7**) that extends the entire

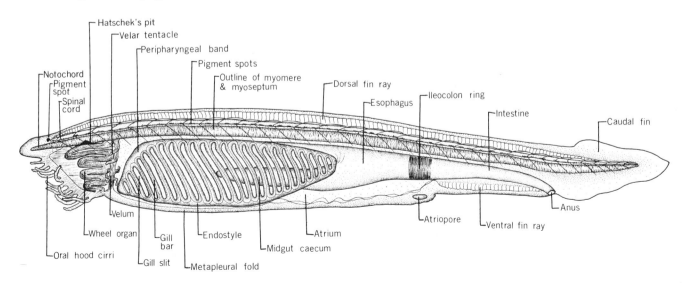

Fig. 1-7. Lateral view of an adult lancelet, showing the major internal features.

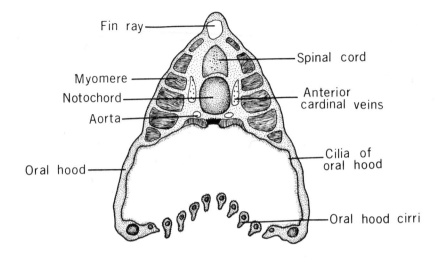

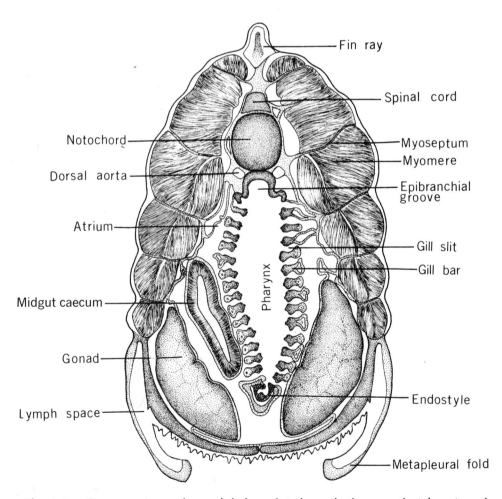

Fig. 1-8. Cross sections of an adult lancelet through the mouth (*above*) and pharynx (*below*).

length of the body immediately beneath the dorsal tubular spinal cord.[1] As an aberrant feature the notochord extends well into the anterior head region, probably as a burrowing adaptation. There is a series of metamerically arranged nerves along the dorsal tubular nerve cord, but there is no expansion of the nerve cord anteriorly as a "brain," and there are few signs of true sense organs except for a large anterior pigment spot and the smaller pigment spots along the spinal cord, perhaps light-sensitive in function.

Circulation in lancelets generally conforms to the vertebrate pattern. Blood is routed anteriorly in the ventral region, up through the gills, and posteriorly in the dorsal region. However, breathing probably occurs primarily by diffusion across the skin and not in the gill region; there are no respiratory pigments. There is no heart, but the blood is kept flowing by waves of contractions of the major vessels and by the contractions of bulblike enlargements of the vessels below the branchial region. There are veins equivalent in function and general location to the anterior and posterior cardinals, hepatic portal, and hepatic veins of higher vertebrates (to be studied later). Thus, the circulatory pattern is similar to that of vertebrates.

There are no true paired fins in the lancelets; however, there is a pair of ventrolateral folds known as **metapleural folds** (G., *meta*, after, and *pleura*, side), which extend from the anterior region of the pharynx to the **atriopore** (best observed in cross section **Fig. 1-8**), and a median dorsal fin, supported by internal rods of connective tissue (fin rays), which extends

[1] Recent studies using the electron microscope have shown that the fine structure of the notochord differs significantly among the various chordates. Vertebrates and enteropneusts have a notochord consisting of vacuolated fiber-containing cells that are epithelial in character. The lancelet notochord is composed of specialized muscle plates. Most tunicates have a notochord of solid yolk- and glycogen-containing epithelial cells, but notochords of the Larvacea are formed of flat epithelial cells (Welsch and Storch, 1971).

along the entire dorsal length of the body. A caudal fin is also present.

Though the sexes are separate, in both males and females the gonads are numerous, segmentally arranged structures (Fig. 1-6), and in this respect differ from those of other chordates.

The anatomy of the feeding apparatus is of great interest. Note in Fig. 1-7 that anteriorly there is a chamber, the **oral hood**, which has around its periphery a series of small tentacles called the **oral hood cirri**; these are thought to serve a chemoreceptive function, but they also selectively introduce food particles into the **buccal** or **mouth funnel** located deep within the oral hood. Each cirrus is supported by a skeletaginous rod of connective tissue and all are joined at a common base. Within the oral hood are located numerous ciliary bands that serve to sweep water into the mouth cavity, and deep within the cavity are fingerlike projections that guard the opening to the pharynx. These projections constitute the **wheel organ**, whose action forms a current of water that will flow into the pharynx. Thus, water along with minute food particles is funneled into the mouth, first by the cirri of the oral hood and the internally located ciliary bands, and then into the buccal funnel by a current created by the wheel organ. The entrapment of food particles is aided by a mucoid secretion of the most dorsal of the wheel organ's ciliated grooves, known as **Hatschek's pit**. The water thus enters the pharyngeal region, or **branchial basket**, and is acted upon in much the same manner as in the tunicates. Food particles are filtered out of the water by the gill bars of the branchial basket, and the water passes into the atrial chamber surrounding the branchial basket and thence out its own posterior exit, the atriopore. Meanwhile, the collection of the food particles is enhanced by a mucoid secretion of a longitudinal midventral hypobranchial groove known as the **endostyle** (**Fig. 1-8**), a groove along the ventral floor of the pharynx considered to be partly homologous to the thy-

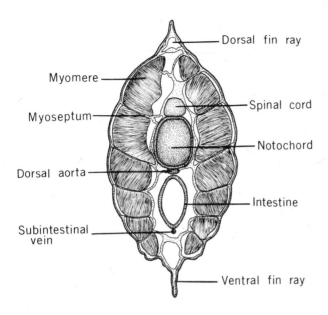

Fig. 1-9. Cross section of an adult lancelet through the tail region.

roid gland of higher vertebrates. Ciliary currents formed by cilia along the gill bars carry the food particles in streamers in a sort of conveyor-belt system dorsal within the pharynx in an **epibranchial groove,** and then posteriorly into the intestine where digestion takes place. In addition to the intestine there is a **midgut caecum,** an anteriorly extending evagination of the intestine into which small food particles are shunted for digestion. Larger particles pass directly down the intestinal tract. A darkly staining band along the alimentary canal, the **iliocolon ring,** is easily observed also, but its function is enigmatic.

In the pharyngeal sections the highly modified coelom of this organism is present but difficult to observe. It should be noted that the term **coelom,** often used synonymously with "body cavity," refers not just to a body cavity, but to body spaces that are bounded entirely by an epithelium of mesodermal origin.

Lancelets differ drastically from the vertebrates in their excretory system. Vertebrates have developed an efficient kidney as both an excretory and water-filtering system; lancelets, like many of the invertebrates, have segmentally arranged nephridia.

We may now ask the question of the phylogenetic position of the lancelets: Where do they stand in relation to the other protochordates and to the vertebrates?

It is instructive to consider the possible derivation of the cephalochordates from ancestors resembling their cousins, the tunicates. The larval form of the tunicates, with few modifications, and remaining in a neotenic state, could easily be the source of a cephalochordate-like organism. The larval forms have a filter-feeding mechanism not unlike that of the cephalochordates, and in the tail region there is a notochord located immediately below a nerve cord. In the tail there is vertebratelike muscle, but no segmentation; the transition would not be difficult.

The numerous similarities between the cephalochordates and the vertebrates cannot be ignored, nor can their resemblance to the "lower" protochordates be dismissed. Though there is still some disagreement among zoologists as to their exact phylogenetic position, most students of vertebrate evolution hold that the cephalochordates are rather specialized organisms that closely resemble, and are derived from, animals on the main line to the vertebrates; in other words, they are close to the mainstream of vertebrate evolution. Without certain specializations, such as an anteriorly extending notochord, a nephridialike excretory system, and other minor features, they would certainly be "good" vertebrate ancestors.

It should be pointed out here that the neural crest cells (which bud off the developing neural tube; see Fig. 7-1) are an innovation with the vertebrates (see Fig. 1-1), and are not to be found in the cephalochordates. During ontogeny the neural crest cells migrate over much of the vertebrate body and give rise to or induce the formation of such complex structures as certain cartilages and bones of the head region, spinal and cranial ganglia (particularly of the autonomic nervous system), nerve sheaths, melanophores, part of the dermis, and other important structures. The neural crest is thus of great evolutionary significance in the origin of the vertebrates.

Ammocoetes Larva

Although the adults of the living cyclostomes will be covered in the next chapter, it is instructive to consider now the larval form of the lamprey, the ammocoetes larva. Lampreys typically ascend rivers or streams to spawn during the breeding season, and the resulting ammocoetes larvae, which attain a length of 4 to 5

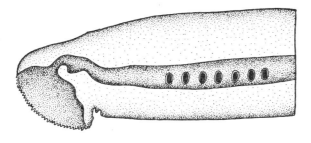

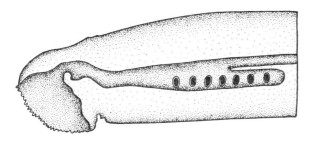

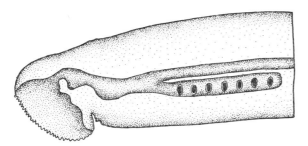

Fig. 1-10. Diagrammatic illustration of a sagittal section of the ammocoetes larva undergoing metamorphosis into the adult lamprey, showing the separation of the esophagus and the pharynx.

inches, remain for the most part buried in the mud or sand for several years (up to seven) with only their heads protruding, much in the same manner as the lancelets. Unlike the adult lamprey, and as in the lancelets, the buccal funnel gives rise to but a single pharyngeal chamber, which at metamorphosis becomes transformed into a separate esophagus and pharynx (**Fig. 1-10**). After metamorphosis the ammocoetes is transformed into an adult lamprey and descends the stream to the marine home of the adult. Some lampreys are confined (secondarily) to the Great Lakes, but most are marine. The ammocoetes larva is of great interest to the comparative anatomist owing to its resemblance to the cephalochordates. We will now discuss the major features of this larva's anatomy (**Figs. 1-11, 1-12**).

As in the cephalochordates, muscle segmentation, or myomerism, in conjunction with an efficient and well-developed notochord, makes possible undulatory swimming movements. The simple V-shaped myomeres of the cephalochordates have in the ammocoetes become transformed into W-shaped structures, for more efficient undulatory swimming. The W-shaped myomeres permit an undulatory rather than unidirectional movement of the body. In addition, the skeleton is much more complex than that found in the cephalochordates. It is composed of cartilage, and consists of a branchial basket in the pharyngeal region and cartilaginous fin supports for the dorsal and caudal fins; these fins are continuous with each other. The notochord is well developed, but unlike that of the cephalochordates, does not extend into the anterior head region, but terminates beneath the posterior region of the brain. The spinal cord is expanded anteriorly as the brain, which may be subdivided into three regions, the hindbrain (**rhombencephalon**), the midbrain (**mesencephalon**), and the forebrain (**prosencephalon**). The primordia of the lateral eyes are apparent as dark spots between the mesencephalon and prosencephalon, and the inner ear primordium is present posteriorly at the area of

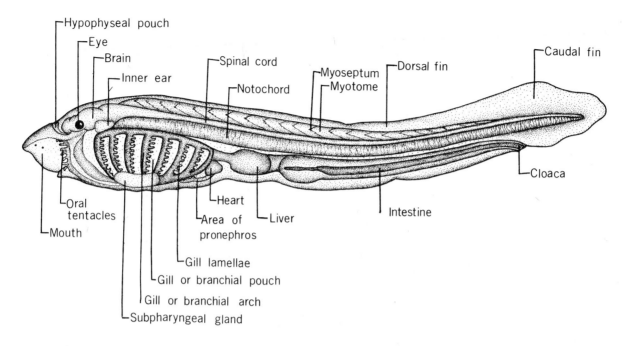

Fig. 1-11. Lateral view of an ammocoetes larva, showing the major internal features.

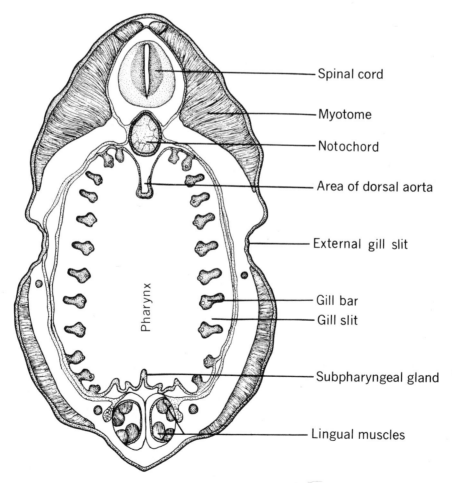

Fig. 1-12. Cross section of an ammocoetes larva through the pharynx.

the rhombencephalon. Anterior to the brain is a small protuberance, the **median nostril,** which leads to the single dorsomedial **hypophyseal pouch.**

Water enters the mouth in much the same manner as in the cephalochordates, and is shunted into the pharynx with the aid of oral tentacles and the velum, not easily observed in specimens. However, the current for feeding is caused by the muscular action of the velum and pharynx rather than by ciliary action as in the cephalochordates. Once water enters the pharynx, the food particles are entrapped in the **gill lamellae** with the aid of a mucoid secretion of the **subpharyngeal gland** or **endostyle,** which upon metamorphosis gives rise to the thyroid gland of the adult. The food pellets pass posteriorly to the esophagus and intestine, while the water passes out the seven **external gill slits.** The **liver** is located at the junction of the esophagus and intestine, and the ventral **heart** is present anterior to the liver. The circulatory pattern is generally similar to that of the lancelets, but blood cells and hemoglobin are present and gaseous exchange takes place across the gill epithelium. Thus the ammocoetes pharynx exhibits the two primitive primary functions characteristic of the chordate pharynx—food-getting and gaseous exchange. The area of the **pronephric kidney** is slightly anterior to the heart.

It cannot be denied that the ammocoetes larva is of great phylogenetic interest. So different is it from the adult that it was first called by a distinct generic name, *Ammocoetes,* and was only later discovered to be the larval form of the lamprey. Its superficial resemblance to the lancelets is remarkable; it is a filter-feeder, but lacks the atrium, and the water exits via the external gill slits. However, it has all the basic vertebrate morphological features. The ammocoetes exhibits overt cephalization; there is a well-developed brain, and associated sense organs. It has more-advanced myomeres than the cephalochordates, permitting complex undulatory movement. In addition, it uses mus-

cular action to aid in feeding rather than simple ciliary movement, and the gill pouches take on their primitive vertebrate function as a system for both feeding and gaseous exchange.

In **Fig. 1-13** are illustrated the major features thought to have characterized the origin of the primitive chordates and vertebrates. In summary, the echinoderms may owe their ancestry to forms not drastically unlike the living pterobranchs, at least superficially. These forms would have presumably been primitive sessile arm-feeders. The acorn worms are hypothetically derivable from pterobranch descendants with a gill filter-feeding mechanism. The tunicates may likewise represent descendants in which the adult filter-feeding mechanism was stressed; but in tunicates we see the development of free-swimming larvae with a notochord, dorsal nerve cord, and well-developed swimming adaptations. With the abandonment of the sessile adult form through the evolutionary process of neoteny, and the increased emphasis on the evolution of swimming adaptations, the evolution of lanceletlike forms would have begun. Perhaps the ammocoetes larva is the next living "missing link" in vertebrate origins.

Tunicate and Acorn Worm: External Features

Familiarize yourself with the anatomy of a tunicate (Fig. 1-5), and obtain a preserved adult specimen for examination of the external features. The specimen should be placed in a finger bowl for examination. Examine the incurrent and excurrent siphons. Water enters the animal through the incurrent siphon and exits through the excurrent siphon. Examine the external covering. It is called a tunic, and contains a substance, tunicin, which is somewhat similar to cellulose. The internal anatomy is difficult to observe; however, if demonstrations

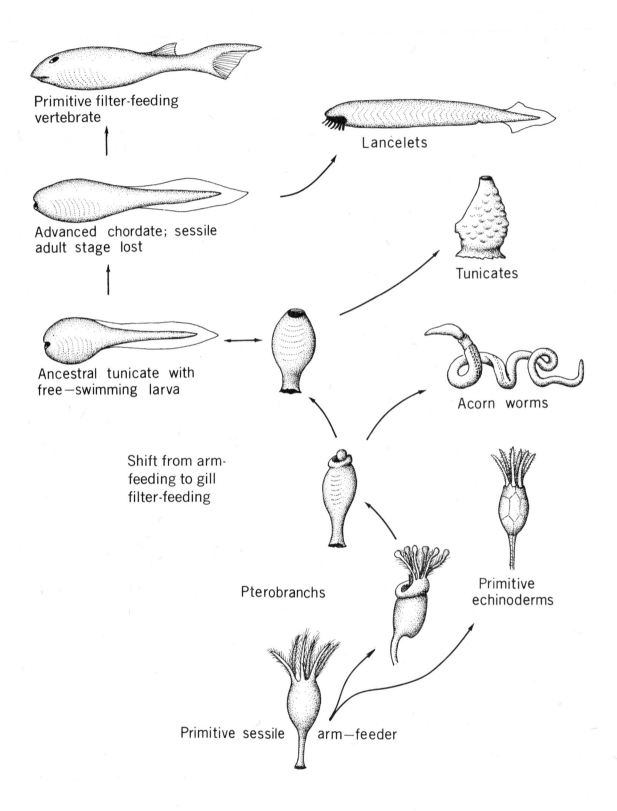

Primitive filter-feeding
vertebrate

Advanced chordate; sessile
adult stage lost

Ancestral tunicate with
free—swimming larva

Lancelets

Tunicates

Acorn worms

Shift from arm-
feeding to gill
filter-feeding

Pterobranchs

Primitive
echinoderms

Primitive sessile arm—feeder

Fig. 1-13. A hypothetical representation of the major features thought to have characterized the origin of the vertebrates. (After Romer, 1970.)

are available, observe the major features of the branchial basket (pharynx) and the digestive system illustrated in Fig. 1-5.

Obtain a preserved specimen of an acorn worm and study its external features by referring to Fig. 1-3.

Lancelet: External Features and Cross Sections

Obtain a preserved whole specimen and a stained microscope slide of a lancelet and study the anatomical features illustrated in Fig. 1-6 and 1-7. The preserved whole specimens clearly illustrate the myomeres, myosepta, and fins. There is a dorsal fin that extends along the top, and a ventral fin in the posteroventral quarter of the organism. Observe the metapleural folds that continue anteriorly and laterally on either side as continuations of the ventral fin.

Using the stained microscope slide, observe the same features you observed on the preserved specimens. Begin anteriorly, and study the structure of the mouth: the oral hood, oral hood cirri, wheel organ, Hatschek's pit, velum, and velar tentacles. Note the position and extent of the notochord and its relation to the spinal cord. A large pigment spot may be observed anterior to the spinal cord, and numerous smaller spots along the entire area of the spinal cord. Examine the numerous gill bars of the branchial basket, and locate the endostyle in the midventral region of the pharynx. Study the following features posteriorly: midgut caecum, esophagus, ileocolon ring, intestine, anus, atrium, and atriopore; and determine the functional relationship of the different features of the pharynx and digestive regions.

Obtain a stained microscope slide of cross sections of a lancelet at the levels of the head, pharynx, and intestine, and study the anatomical features illustrated in Fig. 1-8 and 1-9. In the cross section at the level of the head, locate the fin ray, spinal cord, notochord, myomere, anterior cardinal veins, aorta, and the oral hood (cilia and cirri).

Now turn to the section at the level of the pharynx, and observe the large myomeres with their corresponding myosepta. Easily observed in the dorsal region are the fin ray, spinal cord, notochord, and paired dorsal aortae. Study the relationship of the atrium to the gill bars and gill slits. Note the epibranchial groove dorsally and the endostyle ventrally. The large midgut caecum is easily observed on the right-hand side of the organism. Be sure that you understand precisely how the organism feeds. Located laterally in the ventral region are the large gonads. Testes will appear as bodies exhibiting very small dots, while ovaries exhibit numerous large nucleolated cells; gametes are discharged directly into the atrium, and exit via the atriopore. The metapleural folds are best observed in the pharyngeal sections.

Features that are observable in the intestinal sections are the ventral fin ray, subintestinal vein, intestine, and single median dorsal aorta.

Ammocoetes Larva: External Features and Cross Section

Obtain a stained microscope slide of a whole specimen and a slide of a cross section through the pharynx of the ammocoetes larva, and study the anatomical features illustrated in Figs. 1-11 and 1-12, comparing them with equivalent structures in the lancelet. Observe the myomeres and myosepta, and the relationship of the notochord to the spinal cord and brain. Note that in this organism, unlike the lancelet, the notochord extends only slightly beneath the expanded area of the anterior spinal cord (brain). Note also the hypophyseal pouch, and the eye and inner ear. Within the mouth cavity are located the large oral tentacles. Ammocoetes larvae and lancelets feed similarly, but in the former the water taken into the pharynx exits via external gill slits, rather than via an atriopore. Examine the pharyngeal region, and locate the gill pouches, gill arches, and gill lamellae. The subpharyngeal gland is homologous to the endostyle of the lancelet and serves a similar function. Note the area of the pronephric kidney, the heart, and the liver. In the posterior region observe the intestine, cloaca, and dorsal and caudal fins.

In cross sections, note the large myotomes (embryonic myomeres), and the relation of the spinal cord to the notochord. The large dorsal aorta is easily observable. In the pharyngeal region, study the functional relationships of the various features—gill bars, gill slits, external gill slits, and subpharyngeal gland—and be sure that you understand how feeding occurs. In the ventral region locate the large lingual muscles.

Observations on a Filter-feeding Vertebrate

Filter-feeding is not confined to the protochordates; such diverse vertebrates as flamingos, certain whales, and the tadpoles of some frogs feed by filtering microscopic particles out of the water. The tadpoles of the African frog genus *Xenopus* feed by filtering microscopic particles from the waters in which they develop. Water is taken into the pharynx via the mouth, and the flow of water is unidirectional through the animal. An elaborate set of gills separated by gill slits is present on both sides of the pharynx. Water passes over the gills and exits from the pharynx through the gill slits, passing into chambers behind the gills. Water leaves the body via two small spiracles at the posterior end of these chambers. Microscopic food particles are trapped in mucus secreted by the endostyle. Abundant cilia line the internal pharyngeal surface. Contractions of the cilia move the mucus as a sheet from the ventrally located endostyle, dorsally across the gills and to the roof of the pharynx. Cilia in the pharyngeal roof shape the food-bearing mucus into a cord and move it into the esophagus. This cord then passes to the gut for digestion. The movement of trapped food particles in the pharynx is easily observed.

Catch one tadpole from the holding aquarium using a dip net, and place the specimen in a petri dish containing a heavy suspension of either activated charcoal or red carmine dye for two to four minutes. Be sure that the petri dishes are clean, as the tadpoles are very susceptible to fixatives, such as alcohol and formalin, and other contaminants. Remove the tadpole from the suspension and place it in a small petri dish containing clear water from the holding aquarium. Place the petri dish under a dissecting scope and illuminate. Be particularly careful not to place the light so close to the specimen as to overheat them. The pharynx will be full of readily visible trapped particles. Study the movements of food in the pharynx and be sure that you understand the principles and mechanics involved.

2

Jawless Fishes

The modern eel-like lampreys and hagfishes (**Fig. 2-1**), vernacularly known as the **cyclostomes** (G., *kyklos*, circle, and *stoma*, mouth), are the only living representatives of the otherwise completely extinct **class Agnatha,** or jawless fishes. This class first appeared in the Ordovician Period (approximately 400 million years ago) as well-developed, heavily armored forms collectively know as the **ostracoderms** (G., *ostracon*, shell, and *derma*, skin). They are of great interest, as they are the first known vertebrates. Ostracoderms had, over much of the body (though especially in the head region), a heavy, dermally derived armor or exoskeleton, but they lacked paired appendages—unless the immovable pectoral spines or flaps of

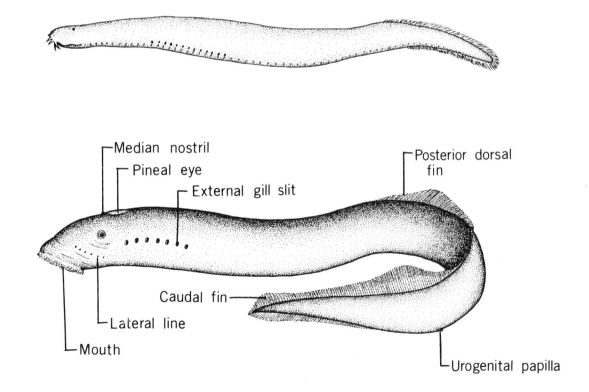

Fig. 2-1. Lateral view of the hagfish, genus *Myxine* (*above*), and the lamprey, genus *Petromyzon* (*below*). (Hagfish, after Dean; lamprey, after Jordan and Everman.)

some forms might be considered as such. In most of these fishes there was a large head region occupied by an expansive branchial area; and many are thought to have made their livelihood by filter-feeding—possibly in much the same manner as some of the protochordates. However, some forms like the cephalaspids (**Fig. 2-2**) were surely bottom-feeders, and it is possible that certain of the heterostracans (Fig. 2-2) were grazers. Of anatomical and evolutionary interest is the fact that many of the ostracoderms had a single median nostril or nasohypophyseal opening in the dorsomedial plane, anterior to a pineal eye. This is known in other vertebrates only in the living cyclostomes.

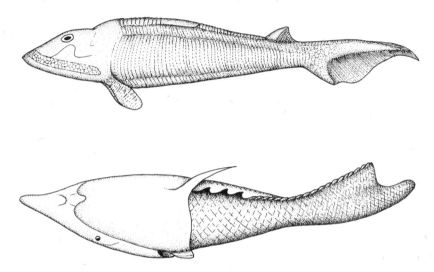

Fig. 2-2. *Hemicyclaspis (above),* a Silurian cephalaspid ostracoderm of the order Osteostraci; *Pteraspis (below),* a Silurian ostracoderm of the order Heterostraci, the first order of the Agnatha to appear.

In the evolutionary lineage leading to the cyclostomes, the heavy armor has been lost, and a slimy integument covers the entire body. This is probably an advanced specialization in the group. The lampreys consist of marine and freshwater species, one having become established permanently in the Great Lakes. Lampreys migrate up freshwater streams to spawn; fertilization is external. Lampreys are both free-living and parasitic. Parasitic species attach themselves to other fishes using the suction mouth, and rasp into the host's flesh with the horny tongue. An anticoagulant is secreted, and the blood is then sucked from the host. The hagfishes differ from the lampreys in that they are scavengers; utilizing their horny rasps, they burrow into the flesh of dead or moribund fish. They also differ in that development is direct; there is no larval stage.

Persistence of these structurally primitive vertebrates through time is no doubt largely attributable to their acquisition of a specialized parasitic mode of life. Cyclostome anatomy is of great interest because it probably most closely approximates that of the first vertebrates.

Lamprey Morphology

Obtain a preserved specimen of a lamprey. Orient the specimen using the anatomical terms in Appendix I, and study the external features illustrated in Fig. 2-1. The skin consists of a complex multilayered epidermis that overlies a dense dermal layer, but is separated from the latter by a connective tissue layer containing pigments and capillaries.

In the anteriormost region of the lamprey is the primary diagnostic feature of the agnatho-

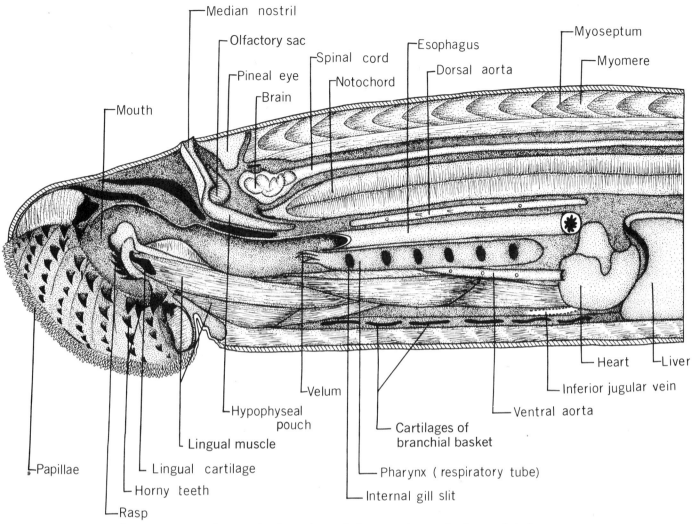

Fig. 2-3. A median sagittal section of the anterior end of an adult lamprey, showing the major internal morphological features.

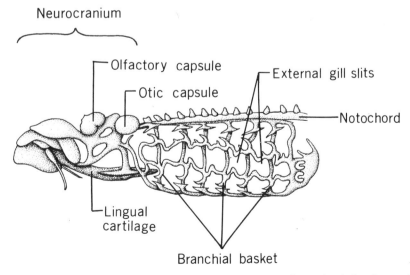

Fig. 2-4. Lateral view of the neurocranium and branchial basket of the lamprey. (Drawn from specimen and following Parker and Haswell, and Kingsley.)

stome fishes, the presence of a mouth without jaws; from this condition the class Agnatha derives its name (G., a-, without, and *gnatha*, jaws). All vertebrates except the class Agnatha are characterized by a mouth with jaws, and may thus be called collectively **gnathostomes** (G., *stoma*, mouth). The mouth of the adult lamprey is a highly specialized structure adapted for a parasitic mode of existence. The lamprey uses its epidermally derived **keratinized teeth,** which differ from the true teeth of other vertebrates, to attach itself to the host to suck its blood. Surrounding the mouth is a periphery of sensory papillae that may also aid in forming a suction attachment to the host. Deep within the buccal funnel is the **horny rasp** on the protrusible tongue that serves to scrape a hole in the host through which blood is procured. The massive tongue muscles are best observed in sagittal section (**Fig. 2-3**).

Turning to other features of the lamprey head, there is a series of sensory pores on various regions of the head, constituting the **lateral line canal system** (Fig. 2-1) of the head, which may be called the **cephalic canal system;** it functions in detecting movement or vibrations in the water. Paired eyes are located laterally on either side of the head; and the single dorsomedial **pineal eye** is present just posterior to the dorsomedial **nostril.** A pineal eye occurred in most of the ostracoderms and many other fossil groups, and occurs today in many modern lizards. Though the precise function of the pineal eye in the various groups of vertebrates remains poorly understood, it is thought to function in some capacity regulating certain endocrinological functions.

The lamprey endoskeleton consists of several distinctive cartilages in the head region and a cartilaginous **branchial skeleton** (G., *branchia*, gills) shown in **Fig. 2-4.** The cartilages of the branchial basket or visceral skeleton support the gills. Each **visceral arch** (column between successive gill slits) has a visceral cartilage and striated branchiomeric muscle; thus the gills are supported by a visceral skeleton and

movement is accomplished by a set of striated muscles. The cartilaginous **neurocranium (chondrocranium)** surrounds and protects the brain and associated sense organs.

In addition, the lamprey has a notochord as its primary axial skeleton, but the notochord (as in all other vertebrates) does not extend into the anterior head region as in the cephalochordates. The notochord consists of turgid, vacuolated cells that are surrounded by a sheath of connective tissue. There is a series of metamerically arranged cartilages that are located above the notochord and partially surround the spinal cord laterally along much of its length. These are probably the rudiments of the neural arches seen in higher vertebrates and may be called **arcualia** (L., *arcualis*, arch-shaped). A similar series in the tail region extends ventrally and is similar to the hemal arches of higher vertebrates.

Make a median sagittal section of the lamprey beginning at the anterior end and extending to approximately 2 inches posterior to the last external gill slit; be sure that your incision passes precisely through the middorsal and midventral lines. Using another specimen, make the following cross sections. First, make a cross section immediately posterior to the median nostril, cutting ventrally, transecting the lateral eyes. Second, make a cross section approximately halfway between the last external gill slit and the cloacal opening. Study the morphological features illustrated in **Figs. 2-3, 2-5,** and **2-6,** and using a probe, determine the relationships of the various structures.

First, examine the gill region in sagittal and cross section. There are seven **gill pouches,** each of which has both internal and external gill openings. Each unit of the gill region consists of an anterior highly vascularized series of **gill lamellae** (demibranch or hemibranch), separated from a posterior demibranch by a gill slit. Successive gill pouches are separated by membranous **interbranchial septa.** An anterior demibranch and a posterior demibranch are known as a **holobranch.** On either side of the

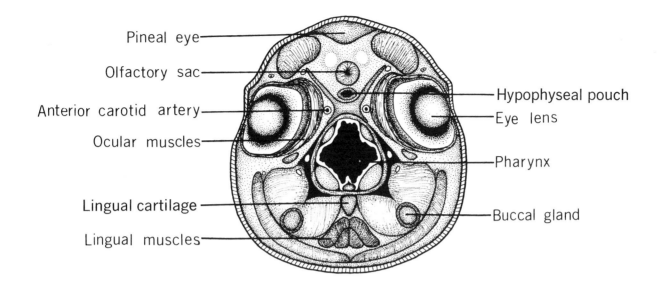

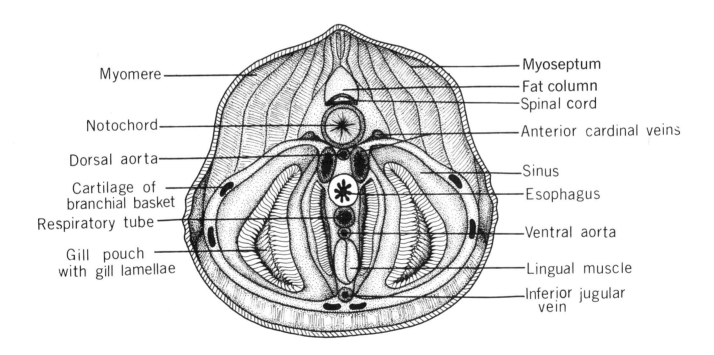

Fig. 2-5. Cross section through the eyes (*above*) and through a gill pouch (*below*) of an adult lamprey.

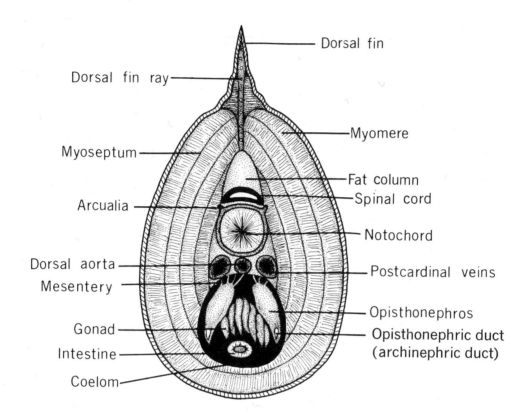

Fig. 2-6. Cross section of an adult lamprey (male) at the level of the first dorsal fin.

interbranchial septa are **lymph sinuses** that contain a blackish lymph.

The **esophagus** and **respiratory tube** are separate. Of special interest here is the **olfactory sac,** the epithelium of which has a greatly increased surface area owing to the infoldings of tissue. Water is continually pumped in and out of the median nostril and into the prenasal sinus by the movements of the pharynx; thus, a continual flow of water over the olfactory organ enables the organism to be continually aware of its chemo-environment. Blood enters the mouth and is shunted into the esophagus with the aid of the **velum,** a flaplike structure composed of fingerlike projections, which guards the entrance to the esophagus. The velum is at the anterior end of the pharyngeal region; thus, when the animal is feeding, blood may be prevented from entering the respiratory system, and water is pumped in and out of the seven **external gill slits.** The pharyngeal tube is a

blind pouch, but the esophagus gives rise posteriorly to the **intestine,** which continues on to the **anus.**

The **brain** and **spinal cord** are best observed in the sagittal section (Fig. 2-3). The nervous system consists of a spinal cord with alternating dorsal and ventral roots of spinal nerves. The roots do not unite to form a common nerve as in the higher vertebrates. The lamprey is presumed to be the primitive state for vertebrates. The spinal cord is expanded anteriorly as a brain, consisting basically (as in the ammocoetes larva) of three major areas. As in other fishes, ten cranial nerves are present.

As a mechanism to maintain equilibrium, an inner ear is present as in other fishes. It consists of the **membranous labyrinth** with **semicircular canals;** these will be discussed in a later chapter.

The mesodermal epithelium surrounding the body cavity or forming the wall of the body cavity is known as the **parietal peritoneum;** that surrounding the coelomic viscera, the **visceral peritoneum.** Various mesenteries suspend the coelomic viscera; the **mesorchium** suspends the testes and the **mesovarium** suspends the ovaries. In each animal there is but a single gonad; ovaries may be distinguished from testes because they are larger and contain numerous granules that are the prospective ova. The ripe gametes, eggs or sperm, escape first into the coelom, and find their way to the exterior via **genital pores** located on each side of the **urogenital papilla.** In the midventral line in the posterior end of the lamprey there is a single exterior opening, the **cloaca,** which is the receptacle for waste products exiting from the intestine via the anus; the anus is the anterior opening within the cloaca. Posterior to the anus is the urogenital papilla, a fingerlike projection within the cloaca that serves as the exit for urinary waste products from the kidneys.

The single-lobed **liver** of the lamprey is a large structure located in the anterior section of the coelomic cavity, and in its anteriormost extent it is concave and abuts the **pericardial cavity.** The intestine is situated along the ventral section of the gonad; there is an infolding into the lumen of the intestine called the typhlosole, which greatly increases the overall absorptive area.

A pair of **opisthonephric kidneys** (opisthonephroi) appear as large flaps, lateral to the other viscera, and occupy the lateral portion of the coelom. Each of the kidneys is drained by an **opisthonephric duct,** best seen in cross section on the ventral edge of each flap.

The general pattern of circulation is **much** like that of the higher fishes and is not drastically unlike that of the cephalochordates. Blood courses forward ventrally from the **heart** as it is pumped out of the muscular **ventricle** into the **ventral aorta.** This unoxygenated blood passes upward into the capillary beds of the gill lamellae where gaseous exchange occurs. The oxygenated blood may pass then to the head or posteriorly into the **dorsal aorta,** which carries oxygenated blood to the viscera and other parts of the body. Unoxygenated blood is collected by various venous channels and eventually finds its way to the **sinus venosus** of the heart. From there blood goes to the single **atrium** and then to the muscular ventricle, where the circulatory pathway begins anew.

It is in the anatomy of the lamprey that the basic vertebrate blueprint is seen, perhaps more clearly than in any other living vertebrate.

3

Evolution of Gnathostomes

Before beginning a study of the evolution of the various organ systems in vertebrates, it is desirable to present a brief sketch of the evolution of the gnathostome vertebrates. This chapter is an outline of the major features of vertebrate evolution. Representative specimens of the different vertebrate classes should be available in the laboratory for examination. A phylogenetic tree of the major fish groups is illustrated in **Fig. 3-1**, and of the major tetrapod groups, in **Fig. 3-2**.

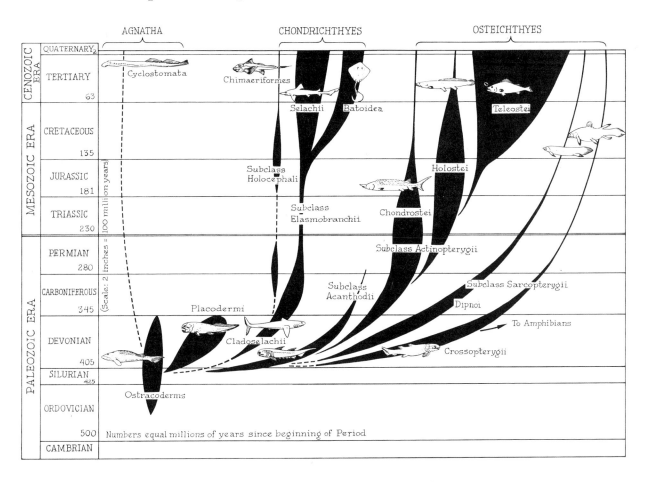

Fig. 3-1. A phylogenetic tree of fishes showing their distribution and relative abundance in time. (From Villee, Walker, and Smith; partly after Romer and Colbert. Courtesy W. B. Saunders Company, Philadelphia.)

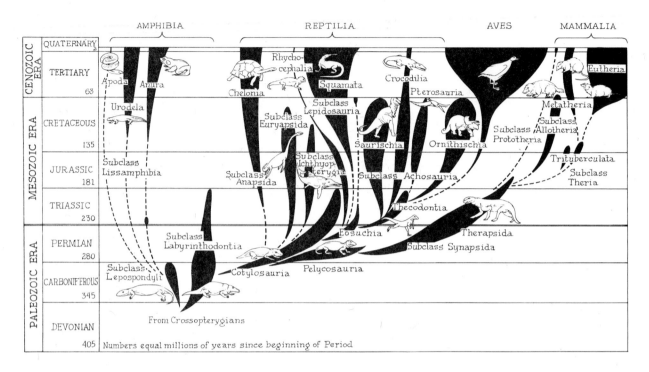

Fig. 3-2. A phylogenetic tree of terrestrial vertebrates. (From Villee, Walker, and Smith. Courtesy W. B. Saunders Company, Philadelphia.)

Class Placodermi (Archaic Jawed Fishes)

At the peak of ostracoderm evolution (toward the end of the Silurian Period) there appeared a group of heavily armored fishes, the **placoderms**, (G, *plax*, plate, and *derma*, skin), which showed advances over the ostracoderms in the evolution of paired appendages, and in the development of movable jaws from an anterior pair of gill arches (see Fig. 4-7). A general term, **gnathostome** (jaw-mouthed), is often applied to the placoderms and all of the higher groups of vertebrates; the ostracoderms and cyclostomes may thus be called **agnathostome** vertebrates. The placoderms were active predaceous animals (**Fig. 3-3**) and their evolution no doubt contributed to the complete extinction of the jawless ostracoderms. Placoderms persisted up to the Permian Period, gave rise independently to two classes, Chondrichthyes and Osteichthyes, and are the only class of vertebrates to have become completely extinct.

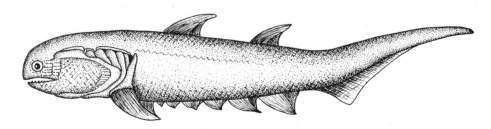

Fig. 3-3. *Climatius*, a Devonian acanthodian placoderm.

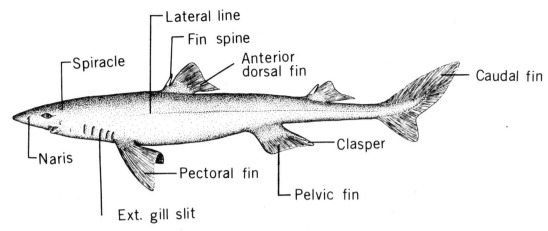

Fig. 3-4. The dogfish, *Squalus acanthias*. The spiracle is the reduced first gill slit of the gnathostome vertebrates; it serves in *Squalus* as an incurrent water passageway. There are five true external gill slits. The lateral line canal system (found in all fishes) is a sensory system for detecting movements in the water. The clasper is a copulatory organ that characterizes male elasmobranchs. The tail or caudal fin is of the **heterocercal** type; the body axis passes upward into the larger dorsal lobe. Fin spines are modified placoid scales.

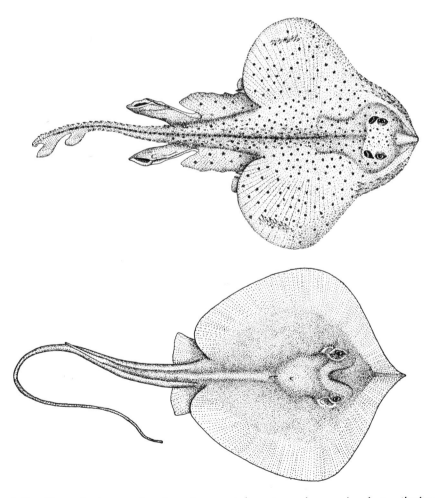

Fig. 3-5. Dorsal views of a female ray (*above*), and a male skate (*below*). The pectoral fins are tremendously expanded. In these bottom-dwellers most of the water enters through the spiracle located posterior to the eyes; the external gill slits are located ventrally. (After Jordan and Everman.)

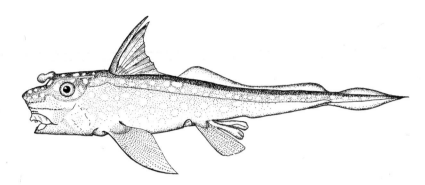

Fig. 3-6. A male ratfish or chimaera. The skin is devoid of placoid scales and the gill slits exit by a single operculum. (After Jordan and Everman.)

Class Chondrichthyes (Cartilaginous Fishes)

The cartilaginous fishes are a specialized group descended from the placoderms, and are characterized by having a skeleton formed entirely of cartilage. No bone is present in any member of the order, though the cartilage is often hardened by calcification. This order originated in the Devonian Period, and has persisted as a very successful group of vertebrates, now consisting of two subclasses. One of these, the **Elasmobranchii,** comprises the order **Selachii,** the sharks (**Fig. 3-4**), and the order **Batoidea,** the skates and rays (**Fig. 3-5**). The other subclass, the **Holocephali,** is composed of a single order, **Chimaeriformes,** the ratfishes or chimaeras (**Fig. 3-6**)—which, unlike the elasmobranchs, have a skin devoid of scales and

an opercular flap covering the gills. The elasmobranchs have a unique type of scale covering most of the body, the **placoid scale** (**Fig. 3-7**), which becomes transformed into fin spines (**Fig. 3-4**), hooks, claspers (**Figs. 3-4** and **3-5**), and stings on the tails of certain rays (**Fig. 3-5**), and even gives rise to the various types of teeth found within the elasmobranchs. Minute placoid scales cover most of the surface of sharks. Each scale consists of a basal plate with a spine projecting posteriorly. This gives the surface a composition like that of fine sandpaper. The cartilaginous fishes were an evolutionary dead end, but have persisted as an important element in the marine environment. Some sharks and skates may be found in freshwater situations.

Class Osteichthyes (Bony Fishes)

The **class Osteichthyes** (G., *osteon*, bone, and *ichthos*, fish) is characterized by a predominantly bony skeleton, though in many species cartilage may persist in the adult. This group, which today comprises more living species (approximately 20,000) than all other classes of living vertebrates, first appeared in the Devonian Period (Age of Fishes), as a group descended from the placoderms, and during the same period split into two groups that are given the taxonomic rank of subclasses. There are the **subclasses Actinopterygii** (G.,

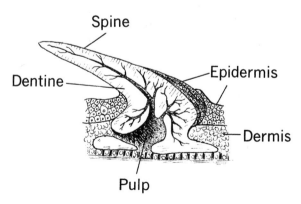

Fig. 3-7. Sagittal section of a placoid scale (dermal denticle) of a dogfish. (After Hertwig.)

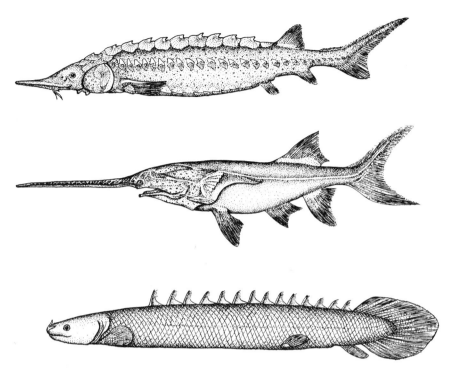

Fig. 3-8. The living chondrostean fishes. Sturgeon (*above*) and the paddlefish (*middle*) are both somewhat degenerate; both have highly modified skeletons and have lost the ganoid scales of their ancestors. The sturgeon has an armor covering of bony plates; the paddlefish is devoid of scales. However, both still possess the heterocercal tail. The Nile bichir (*bottom*), though also modified, resembles more closely the ancient paleoniscoids in its skeleton, and in its true ganoid scales. The sturgeon and paddlefish are foodstrainers. The bichir is a carnivore. (Upper two, after Jordan and Everman; lower, after Gunther.)

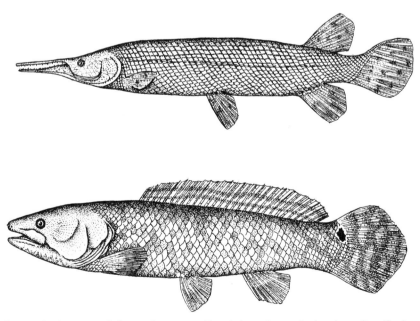

Fig. 3-9. The living holostean fishes, the gar pike (*above*), and the bowfin (*below*). Both are North American freshwater forms. (After Jordan and Everman.)

actino, ray, and *pterygion*, little wing), the ray-finned fishes, and **Sarcopterygii** (G., *sarx*, flesh), the lobe-finned fishes. The actinopterygian fishes are generally characterized by their fanlike fins supported by bony "rays," and in addition lack internal nares. This group was first represented by the **paleoniscoid** fishes, archaic bony fishes with rhomboid-shaped ganoid scales; the paleoniscoids have persisted as the **subclass Chondrostei,** with several "living fossils" among present-day fauna. These include the Nile bichir (*Polypterus*), the paddlefish (*Polyodon*), and the sturgeon (*Scaphyrhynchus*); all are shown in **Fig. 3-8.** A later subclass of ray-finned fishes evolved during the Mesozoic, the **subclass Holostei.** Holostean fishes persist in the modern fauna as two "living fossils," the gar pike (*Lepidosteus*) and the bowfin (*Amia*), **Fig. 3-9.** The holosteans finally gave way in the Cenozoic Era to the bony fishes that characterize the modern fish fauna, the **subclass Teleostei** (**Fig. 3-10**). The

Fig. 3-10. Representative teleostean fishes: a porcupine fish (*above*), a flying fish (*middle*), and a catfish (*bottom*). The upper two are marine teleosts, the lower, freshwater. (After Jordan and Everman.)

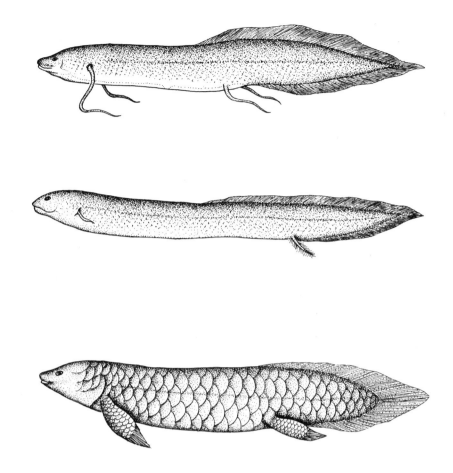

Fig. 3-11. The three living lungfishes (Dipnoi): African (*top*), South American (*middle*), and Australian (*bottom*). All are primary freshwater fishes (they cannot tolerate any salinity), and represent a group that evolved when the major continents were together; hence their present-day distribution. The Australian lungfish (*Epiceratodus*) is the most primitive and resembles the ancient lungfishes. (After Norman.)

lobe-finned sarcopterygians, which are characterized by their fleshly, lobate fins, internal nares, and lungs, were represented in the Devonian Period by the **order Dipnoi** (G., *di*, double, and *pnein*, to breathe), the lungfishes (**Fig. 3-11**). They were once abundant, but now are represented by three living genera, one each in South America, Australia, and Africa. The **order Crossopterygii**, was the most abundant bony fish group of the Devonian. The main line of the crossopterygians, known as the rhipidistians (**suborder Rhipidistii, Fig. 3-12**), was represented by predaceous freshwater forms, and was destined to be the ancestral line leading to the amphibians. A side branch of the crossopterygians, the **coelacanths** (**suborder Coelacanthi**), was thought to be completely extinct, but in 1939 a peculiar fish caught off the coast of South Africa proved to be a living coelacanth—it was called *Latimeria* (**Fig. 3-12**).

Class Amphibia (Amphibians)

The amphibians (G., *amphi*, double, and *bios*, life) were the first group of vertebrates to explore the terrestrial environment. The first of the amphibians, the **labyrinthodonts** (so called because of the complex labyrinthine infolding of the enamel surface of the teeth), are clearly derived from the rhipidistian crossopterygians. Because the crossopterygians had internal nares

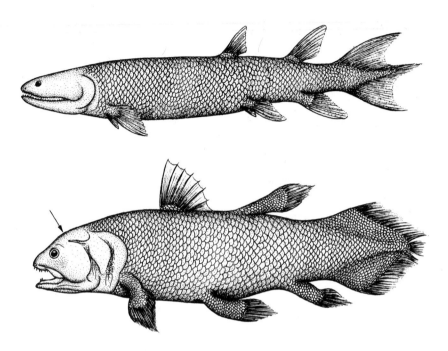

Fig. 3-12. Crossopterygian fishes. *Eusthenopteron* (above) was a typical Devonian rhipidistian, on the main line to the evolution of tetrapods. *Latimeria* (below) is the only known living coelacanth; the arrow shows the position of the spiracle. (*Latimeria,* after Smith.)

and lungs, and fleshy lobe fins with the precursory skeletal elements of the tetrapod limb, the structural changes necessary for the transition to land were probably minor. The amphibians first appeared during Devonian times as the **ichthyostegids** (Fig. 3-13), underwent much of their adaptive radiation during the Carboniferous Period (a time of predominance of large freshwater swamps and warm climates), and gave rise to the reptiles by later Carboniferous time. Though it remains uncertain precisely which of the ancient amphibian groups gave

Fig. 3-13. *Ichthyostega,* among the first of the amphibians, exhibited a strange combination of fish and amphibian characters. The fin rays characteristic of the fish tail were retained in the caudal region.

rise to the modern amphibian fauna, the living Amphibia today consist of several related groups: the frogs and toads (**order Anura**), the salamanders (**order Urodela, Fig. 3-14**), and the legless wormlike **caecilians** of the Old and New World tropics (**order Apoda** or **Gymnophiona**). The amphibians are predominantly freshwater forms that wander out onto the terrestrial environment but must return to water to deposit their eggs. Many forms have an aquatic larval stage, typically a tadpole, which metamorphoses into the adult form.

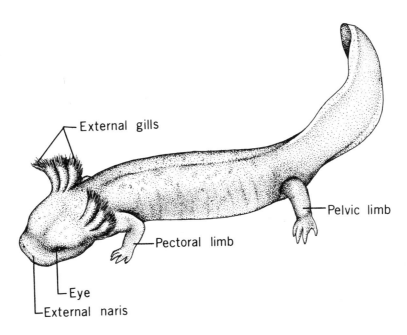

External gills

Pelvic limb

Pectoral limb

Eye

External naris

Fig. 3-14. The mud puppy, *Necturus,* a living urodele. *Necturus* is neotenic; it attains sexual maturity in the larval stage, and never completes its metamorphosis. Because of this, the external gills remain, in addition to the lungs. A lateral line canal system also persists.

Class Reptilia (Reptiles)

Though the first "stem" reptiles (**cotylosaurs**) showed little structural advance over the ancient groups of the amphibians, the reptiles were nonetheless able to capitalize upon their invention of a new mode of reproduction and egg formation, which allowed them to make the full transition to a terrestrial existence without having to return to water to deposit their eggs. Amphibians typically fertilize their water-dependent eggs externally, but reptiles are internal fertilizers (and hence lay fewer eggs), and have a special advanced type of egg (called the **amniotic egg**), which, with its extraembryonic membranes and tough, leathery covering, can develop on land. This advancement literally released the reptiles from the aquatic environment, and allowed a great terrestrial adaptive radiation of vertebrates for the first time.

So successful is the amniotic egg that it has been retained in the same basic anatomical form in all reptiles, birds, and mammals, which are collectively called the **amniotes**. The major features of the amniotic egg are illustrated in **Fig. 3-15**. There are four membranes in addition to the shell. In reptiles and birds the **yolk sac** encloses an abundant supply of nutritive yolk for the developing embryo, whereas mammalian embryos receive nutrients from the maternal blood across the placenta. All amniotes have the following well-developed membranes: The **amnion** encloses the embryo in a

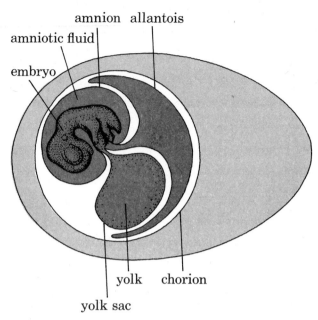

amnion allantois

amniotic fluid

embryo

yolk   chorion

yolk sac

Fig. 3-15. The extraembryonic membranes in a bird's egg. The space between the chorion and the shell is filled with albumen. (From Keeton, Fig. 9.28.)

fluid-filled chamber, thus allowing development in a aquatic medium. The **allantois** serves as a receptacle for the embryo's urinary waste products, and develops blood vessels that, in close proximity to the shell, serve as a breathing organ. The **chorion** is the external membrane surrounding the embryo and the other extraembryonic membranes. In mammals the amniotic egg, lacking a shell, is implanted into the wall of the uterus, where development takes place, and the young are born alive.

The Mesozoic adaptive radiation of reptiles resulted in many distinct groups: some forms returned to freshwater and marine environments (plesiosaurs, ichthyosaurs, chelonians or turtles, crocodiles, and phytosaurs); some became volant (pterosaurs, and some gliding lizards, both extinct and living); some became burrowers (many snakes and lizards); and during the Age of Reptiles (Mesozoic Era), some evolved into herbivorous or carnivorous dinosaurs, which, along with the crocodilians and pterosaurs (all derived from thecodonts) are collectively called the **archosaurs.** The only reptilian groups present in the modern fauna

are the turtles (**order Chelonia**), the **tuatara** or *Sphenodon* (**order Rhynchocephalia**—thought to be on the line ancestral to the living lizards), and the snakes and lizards (**order Squamata**), which persist in both the Old and New World as very successful components of the modern fauna.

Class Aves (Birds)

The evolutionary lineage leading to the birds is perhaps best thought of as just another adaptive line from the great Mesozoic reptilian radiation of the thecodont reptiles. The first bird, *Archaeopteryx,* from sediments of the Jurassic Period of Germany, is one of the most interesting of all fossils. Were it not for the presence of feathers (the most diagnostic avian feature) *Archaeopteryx* would surely have been classified with the reptiles. This crow-sized creature had feathers, but also a reptilian skull with teeth, and claws on its fingers; it was in essence half reptile, half bird. The physiological demands of flight are great, and thus the evolution of birds has been characterized by the acquisition of adaptations associated with flight. The high level of metabolism required by flight is associated with warm-bloodedness (endothermy). Endotherms derive their major heat source from within the body, primarily from muscular thermogenesis; ectotherms (cold-blooded organisms) derive theirs from without. Birds and mammals comprise most endotherms, but certain other vertebrates, such as pythons, some lizards, and even tuna, are capable of considerable endogenous heat production, and may attain endothermy for limited periods.

Such avian features as loss and fusion of bones, pneumatization of bones, loss of teeth, and in the female of most species, retention of only one gonad, are all features that serve to lighten the body for flight. The acquisition of feathers as a modification of the epidermally derived reptilian scales served both as an ad-

aptation for flight, and also for insulation associated with an endothermic existence. Modern birds lay a typical reptilian amniotic egg, but with a highly calcified shell, and have evolved an extraordinary degree of parental care.

Class Mammalia (Mammals)

One line of evolution leading from the "stem" reptiles (**order Cotylosauria**) gave rise to several rather bizarre groups of reptiles, the pelycosaurs, but this same lineage was to culminate in a group known as the **mammal-like reptiles (order Therapsida)**. Therapsids (**Fig. 3-16**) became very abundant in late Permian and early Triassic time, and finally abandoned their reptilian mode of life to become the first mammals. The transition was characterized primarily by adaptations for a more active mode of life; these included the acquisition of

Fig. 3-16. *Seymouria (above)*, a form intermediate between amphibians and reptiles, from sediments of the lower Permian Period of Texas. *Lycaenops,* an advanced mammal-like reptile from the upper Permian of South Africa.

a diversified dentition, the positioning of the legs up under the body to form an upright posture, and the concomitant attainment of warm-bloodedness (endothermy) in order to maintain a constantly active mode of life. Mammals were probably initially nocturnal, and persisted as a group subdominant to the great archosaurian reptiles throughout the Mesozoic Era. The Mesozoic was an era of extremely moderate subtropical and tropical climates over much of the world, thus permitting the cold-blooded (ectothermic) reptiles to maintain a position of dominance in the vertebrate fauna. However, with the advent of greatly changing climates by they end of the Cretaceous Period, and with the extinction of all of the archosaurs except for the crocodilians, the mammals took over as the dominant group, having evolved endothermy at an earlier time.

The present-day mammals include one very primitive group, the **monotremes** (**subclass Prototheria, order Monotremata**), which includes the duckbilled platypus and spiny ant-eaters of the Australian Region. These forms, though possessing many mammalian characteristics in the form of hair, three auditory ossicles, a mammalian jaw, and mammary glands (without nipples), nonetheless still possess certain reptilian osteological features. They are not as effective temperature-regulators as other modern mammals, and lay a typical reptilian amniotic egg. The monotremes are thus intermediate between the reptilian and mammalian levels of organization and may represent a separate offshoot from the therapsid reptiles.

The other two main groups of mammals are thought to have had a common origin, though differing from each other in many features. The **infraclass Metatheria** (**order Marsupialia**), the **marsupials** or pouched mammals, contains such forms as the opossum, kangaroo, koala bear, and Tasmanian devil. Marsupials evolved primarily in Australia and in South America, though most became extinct on the latter continent. Marsupials have a strange manner of reproduction: they give birth to their young at a very early stage; the young emerge, affix themselves to nipples inside the pouch (marsupium), and undergo much of their embryonic development there. The last of the major groups of living mammals is the **infraclass Eutheria**, the **placental mammals,** which comprises the vast majority of living mammals. Eutherians do not emerge from the uterus until well developed, and are termed placental mammals because of the elaborate nature of the placenta (for gaseous exchange and transfer of maternal nutrients to the embryo). Many orders of modern mammals are thought to be derived from basal stocks emerging from an insectivorelike ancestor (**order Insectivora**), though the precise phylogenies are poorly understood.

4

The Cranial Skeleton

The skeleton of the head and pharyngeal regions in the more advanced tetrapods and fishes is so complex and varied that understanding its evolution is very difficult without first considering its structure in primitive vertebrates. The cranial skeleton of primitive vertebrates is composed of three basic distinguishable regions, which in more advanced forms become welded into a single unit.

The primary component of the skull, appropriately termed the **neurocranium,** surrounds the brain and associated sense organs; this was surely the first component of the skull to appear in the early chordates. The **visceral skeleton** [1] is formed of the skeleton of the jaws and branchial (gill) arches. The **dermatocranium** is composed of the superficial dermal bones of the skull. The elements of the neurocranium and splanchnocranium develop during embryogenesis first as cartilaginous components, and may later remain as cartilage or be replaced by endochondral bones. A cartilaginous neurocranium is frequently termed a **chondrocranium.** In addition, areas of the neuro- or splanchnocranium may be ensheathed by dermal bones. **Endochondral (replacing) bones**

(**Figs. 4-1** and **4-2**) are first preformed in cartilage; **dermal** or **membrane bones** have no cartilaginous precursors, but develop directly within the dermis. The three basic cranial components will be considered in primitive and advanced vertebrates.

The Neurocranium

All vertebrate embryos have two basic centers of cartilage formation (**Fig. 4-3**). There is a part of longitudinal cartilages that develop beneath the forebrain anterior to the notochord; these are the **prechordal plates** or **trabeculae,** which later expand and unite to form the **ethmoid plate.** Paralleling the notochord are the posterior longitudinal cartilage centers, the **parachordal plates,**[2] which likewise expand across the midline to form the **basal plate.** Other centers also begin to form as **olfactory**

---

[1] The term *splanchnocranium* is often used as a synonym for visceral skeleton, but is perhaps more appropriately applied to those components of the visceral skeleton that are incorporated into the actual cranium.

[2] It should be noted that the parachordal cartilages are embryologically derived from sclerotomal mesenchyme in much the same manner as the vertebral column; however, the prechordal cartilages are derived from mesenchyme of neural crest derivation, as is much of the visceral skeleton. This embryological fact argues strongly that the prechordals represent an anterior modified visceral cartilage.

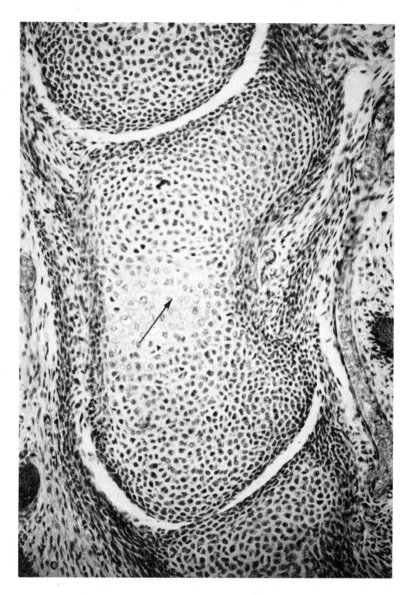

Fig. 4-1. Photomicrograph of a sagittal section of a phalanx (200×) of a hamster embryo, showing the formation of an endochondral bone. The phalanx is composed of cartilage at each end, but in the center an area of endochondral ossification is forming (*arrow*). Cartilage cells (chondrocytes) lie in lacunae and secrete the cartilaginous matrix, which is composed of a glycoprotein (chondromucoid). Cartilage is not penetrated by blood vessels and has no interconnecting canaliculi, so the chondrocytes are supplied by simple diffusion. Cartilage is primarily an embryonic skeletal material that is later replaced by endochondral bone. (Photo courtesy of Drs. Catherine Henley and D. P. Costello.)

capsules to surround and protect the olfactory organs, and as **otic capsules** to surround completely the developing inner ear. In addition, a bridge between the otic capsules is formed posteriorly, dorsal to the brain; it is known as the **synotic tectum.** Both the olfactory and otic capsules fuse with the developing neurocranium, but the cartilage that develops around the eye, the **optic capsule** (which later gives rise to the sclerotic coat of the eyeball), does not; it remains independent to permit eyeball movement.

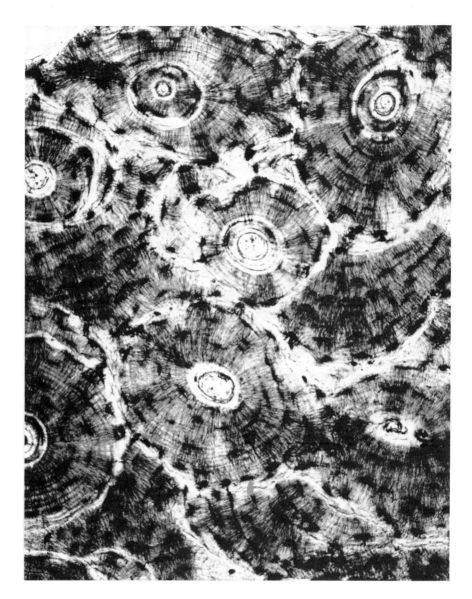

Fig. 4-2. Photomicrograph of a ground section of human bone showing the Haversian systems (200×). Bone cells (osteocytes) become located in small lacunae (black) which are interconnected by small, fluid-filled, concentric canals (canaliculi). This concentric arrangement of the lacunae is termed the Haversian system. The Haversian canals at the center of each Haversian system contain blood vessels. The hydroxyapatite crystals, which are composed of calcium, phosphate, and carbonate, are deposited under the influence of bone cells. In some of the earliest ostracoderms, and in many modern bony fishes, the bone is acellular, a situation in which the osteocytes do not come to lie within the bone, but are displaced as the bone is deposited. Histologically, endochondral and dermal bone are very similar. (Photo courtesy of Drs. Catherine Henley and D. P. Costello.)

Further elaboration of the neurocranium usually involves a more complete encapsulation of the brain along its sides. In most vertebrates the cartilaginous chondrocranium is eventually replaced during embryogenesis by endochondral bone. The dorsal aspect of the brain is protected by a series of dermal bones not associated with the neurocranium. However, the

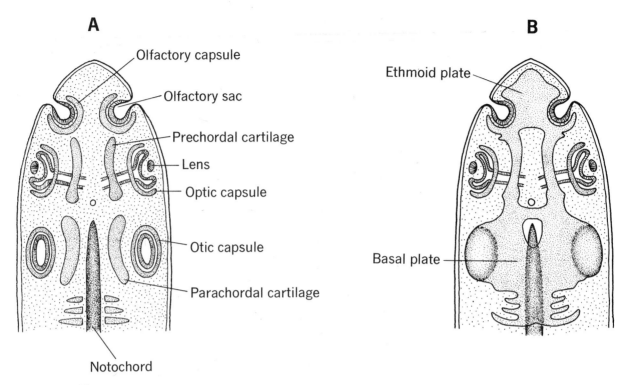

A

Olfactory capsule

Olfactory sac

Prechordal cartilage

Lens

Optic capsule

Otic capsule

Parachordal cartilage

Notochord

B

Ethmoid plate

Basal plate

Fig. 4-3. Diagrammatic illustrations of the ventral aspect of the developing neurocranium in its initial stages (*A*, early; *B*, later).

cyclostomes (see Fig. 2-4) and the chondrichthyian fishes (see Figs. 4-4 and 4-5) have a cartilaginous structure that surrounds the brain both ventrally and dorsally; furthermore, cartilage remains in the adults as the sole skeletal material—there is no bone present in any of these species. In many sharks, however, the cranial skeleton is strengthened by the deposition of calcium salts, forming calcified cartilage. Fossil neurocrania of sharks are known that have been preserved because they are composed of calcified cartilage.

SQUALUS

Obtain a specimen of the neurocranium of *Squalus* for study.

Although the adult neurocranium of the elasmobranchs is a highly specialized structure among the vertebrates, it is convenient to study, and serves as a general guide to the organization of the head. Study the neurocranium of a shark following the description below and **Figs. 4-4 and 4-5.**

The neurocranium may be conveniently divided into a number of regions that will be considered separately.

The nasal region, formed from the olfactory capsules and the anterior prechordals, consists of structures anterior to the antorbital processes. A pair of large **olfactory capsules** house the olfactory organs. The external openings, the **nares,** permit water to flow continually over the olfactory epithelium, and inform the organism of its chemical environment. The anteriorly projecting **rostrum,** which is keeled ventrally and excavated dorsally, supports the snout; the **precerebral cavity** of the rostrum is filled with a gelatinous material in the living animal. **Rostral fenestrae** are present posteriorly.

The orbital region, formed from the posterior prechordals, includes the area between the orbits. Ventrally, the **rostral keel** continues as the **infraorbital ridge,** which merges into a more flattened **basitrabecular region.** Dorsally, the

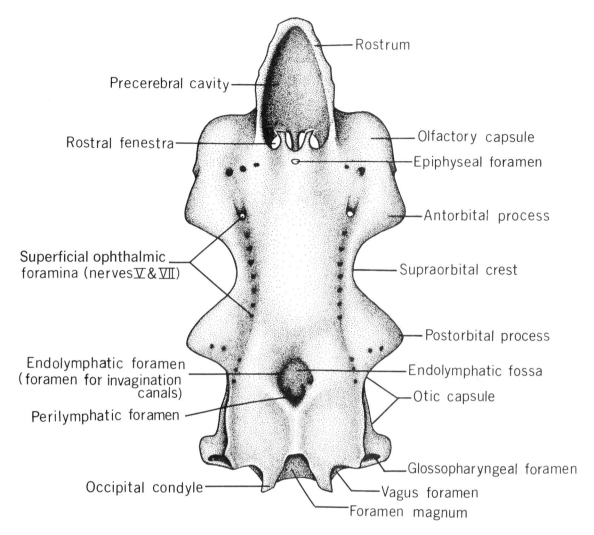

Fig. 4-4. Dorsal view of the neurocranium of *Squalus*.

antorbital and **postorbital processes** border the **supraorbital crest.** Anteriorly, there is a single medial **epiphyseal foramen,** which in life accommodates a vestige of the pineal eye, the epiphyseal stalk. There is also a series of lateral **superficial ophthalmic foramina** for the passage of branches of the superficial ophthalmic nerve (composed of cranial nerves V and VII). In some specimens there will be preserved within the orbit a small structure shaped like a golf tee; it is the **optic pedicel,** which abuts the eyeball for support. In the medial wall of the orbit is a series of foramina for the passage of various cranial nerves and blood vessels. Anteriorly within the orbit is the large **optic foramen** for the passage of the optic nerve; pos-

teriorly, the large **trigeminofacial foramen** transmits the trigeminal (V) and facial (VII) nerves, which are somewhat fused together as they exit the foramen.

Posteroventrally, there is the large **basal plate** formed from the fusion of the parachordal cartilages. It is pierced by the **carotid artery foramen.**

The posterolateral aspect of the neurocranium is formed by the large **otic capsules** that surround and protect the inner ear. Dorsally, there is a medial **endolymphatic fossa,** within which are located two foramina that communicate with the semicircular canals of the inner ear. The anterior **endolymphatic foramina** permit the passage of the invagination canals of

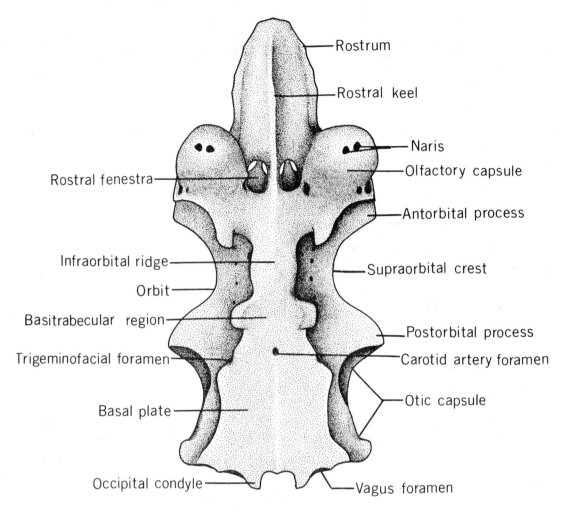

Fig. 4-5.   Ventral view of the neurocranium of *Squalus*.

the inner ear; the posterior **perilymphatic foramina** enter the cavity surrounding the membranous labyrinth (ear).

The occipital region is formed from posterior cartilaginous arches; it surrounds the large **foramen magnum,** which permits passage of the spinal cord. The foramen magnum is bordered laterally by two **occipital condyles** that articulate with the vertebral column. The **glossopharyngeal foramen** and the **vagus foramen** transmit cranial nerves IX and X, respectively.

The neurocranium of most fishes has no roof, except for the area of the synotic tectum, and an occasional bar in the epiphyseal region. However, much of the cartilage of the chondrocranium is replaced by bone (endochondral ossification). Beginning with the most anterior,

there are four major centers of endochondral ossification; ethmoid, sphenoid, otic, and occipital. These bones will be studied in detail only in the mammal.

PRIMITIVE TETRAPODS

In primitive tetrapods the neurocranium did not differ significantly from that found in many fishes; it covered the back, ventral portion, and lateral surfaces of the brain, and was predominantly ossified in the adults.

Necturus is a highly specialized neotenic salamander, and the skeleton is composed of an unusually large amount of cartilage. However, the neurocranium is easily studied and presents some features of the general tetrapod pattern. Study the neurocranium of *Necturus* (**Fig. 4-6**), and identify the anterior **trabecular horns, ethmoid plate, antorbital cartilages,** and **trabeculae.** Note that the **parachordal plates** are modified to accommodate the **fenestra ovalis** of the middle ear (to be studied later), and that certain bones have replaced the cartilage; these include the **prootics, opisthotics,** and **exoccipitals.** The **quadrates** are ossifica-

tions of the posterior ends of the first visceral arch, and will be treated later. The **synotic tectum** joins the two halves of the neurocranium posteriorly.

MAMMALS

The adult mammalian neurocranium is so welded with dermal components that it is best studied along with the dermatocranium.

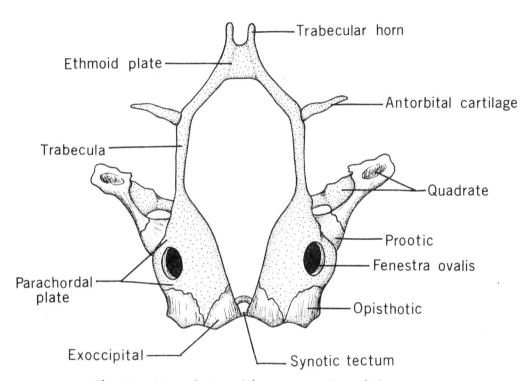

Fig. 4-6.   Ventral view of the neurocranium of *Necturus*.

The Visceral Skeleton

The second feature of the cranial skeleton to be studied is the visceral skeleton (**Fig. 4-7**). This portion of the skeleton consists of the skeletal elements of the visceral arches. These form a series of columns of tissue (primarily of mesodermal origin) in the lateral pharyngeal wall, composed primarily of branchiomeric muscle, with appropriate cranial nerves, aortic arches, and the visceral cartilages or appropriate replacing bones. These cartilages or their replac-

ing bones constitute the actual visceral skeleton. In gnathostomes the first of the visceral arches forms the skeletal support of the upper and lower jaws, and the successive arches follow posteriorly, each separated from the anterior one by a pharyngeal slit (called a gill slit if there is a true gill present). The visceral skeleton is of great historical interest from the standpoint of vertebrate origins. Recall that the vertebrate ancestors were small, sessile, baglike animals, little more than barrel-shaped pharynxes designed for food-sifting. These "visceral" orga-

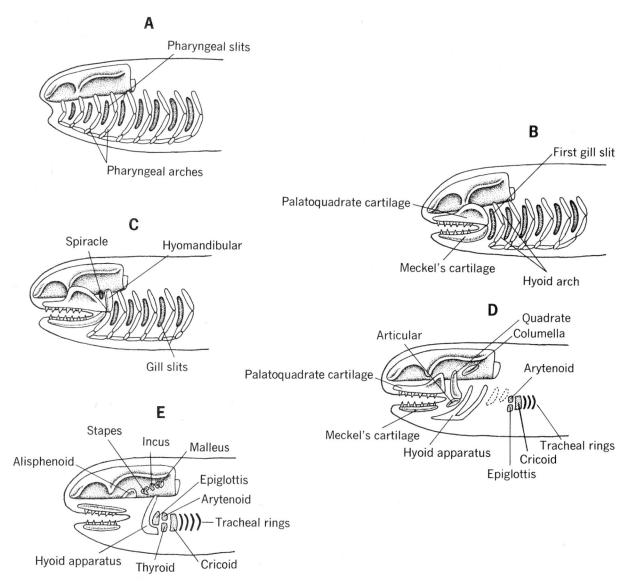

Fig. 4-7. Diagrammatic illustrations of idealized stages in the evolution of the visceral skeleton. *A,* the primitive jawless stage, in which there is a long series of undifferentiated visceral cartilages and a mouth without jaws. *B,* the hypothetical placoderm stage, in which jaws have been formed by an anterior pair of gill arches, the first gill slit is not modified as a spiracle, and the hyomandibular cartilage is not modified as a jaw support. *C,* the structure of most jawed fishes, in which the first gill slit has been reduced to a spiracle owing to the forward migration of the hyomandibular as a jaw-support mechanism. *D,* the advanced amphibian and reptile stage, in which the quadrate and articular bones (formed from the posterior portions of the palatoquadrate and Meckel's cartilages, respectively) constitute the jaw articulation, the hyomandibular cartilage has migrated into the middle ear cavity to form the columella (middle ear ossicle), the hyoid apparatus or glossal (tongue) skeleton is formed from the remainder of the hyoid arch (II), and the posterior visceral cartilages contribute to the cartilages of the laryngotracheal region. *E,* the mammalian visceral skeleton, in which the quadrate and articular bones have migrated into the middle ear to form the incus and malleus, respectively, the alisphenoid bone of the skull is contributed by arch I, and the hyoid apparatus and laryngotracheal cartilages are contributed by the posterior arches.

nisms eventually aquired mobile larval forms for dispersal, but ultimately evolved "somatic" types such as the lancelets. Thus, these organisms made a transition from a predominantly "visceral" to a "somatic" type. Consequently the visceral skeleton, originally a supportive structure for a food-gathering mechanism, eventually became a supporting endoskeleton for the gills—as one sees in the agnathostome fishes—and a modification of one of the anterior pairs became a jaw mechanism.

The visceral skeleton of the cyclostomes has already been treated (see Fig. 2-4). Briefly review its structure. It consists of little more than a series of cartilaginous supports for the gills, and is appropriately termed a branchial basket.

SQUALUS

Chondrichthyian fishes, though highly specialized, exhibit many primitive characters in the visceral skeleton. There are the usual seven visceral arches, with the first modified as a pair of jaws. However, the skeleton is modified in that it has a specialized jaw suspension mechanism (to be discussed later), and lacks a complete gill slit between the jaw arch and the second visceral arch. Some species of placoderms are thought to have had a complete gill slit between the jaws and the next posterior arch. In *Squalus* the second visceral arch (hyoid) has migrated close to the jaws and obliterates much of the second visceral pouch (called a spiracle). In addition, the second arch serves to help support the jaws.

Study a specimen of the visceral skeleton of *Squalus* following the instructions below, and using **Figs. 4-8** and **4-9.** By convention, the visceral arches are numbered by considering the jaws or mandibular arch as number 1.

Note that in *Squalus* there are seven visceral arches, but only five gill slits, there being no gill slit posterior to the last visceral arch. A general tendency in fishes is toward a reduction in the number of gill slits (three is common in

many modern forms), and for all gill slits to exit by a common opening guarded by a bony operculum. Follow the visceral arches from anterior to posterior in your specimen. The first, or **mandibular arch,** consists of the upper and lower jaws. The upper jaw is called the **palatoquadrate cartilage,** the lower, **Meckel's cartilage.** The tooth rows are located on both of these cartilages, but teeth represent part of the dermal skeleton, being modified placoid scales.

There are two small **labial cartilages** attached to Meckel's cartilage. An **orbital process** extends off of the palatoquadrate; it may prevent excessive lateral movement of the jaws. The second visceral arch or **hyoid arch** consists of three cartilages; from dorsal to ventral, they are the **hyomandibular, ceratohyal,** and **basihyal.**

There are many variations in the relationship of the palatoquadrate and the braincase. In elasmobranchs the palatoquadrate is attached loosely to the braincase by ligaments and to a great extent by the orbital process, and the hyomandibular is braced to the otic capsule and buttresses the palatoquadrate; such an arrangement is termed **hyostylic** jaw suspension and is commonly found in many modern teleosts. In situations in which the palatoquadrate is cemented to the neurocranium by dermal bones, the suspension is termed **autostylic.**

The visceral skeleton posterior to the hyoid arch is composed of a series of five multi-jointed cartilages. Each consists of jointed cartilages named, from dorsal to ventral, **pharyngobranchial, epibranchial, ceratobranchial,** and **hypobranchial.** The hypobranchials articulate ventrally with several **basibranchials** that fuse in the midventral line.

Associated with the epibranchial and ceratobranchial cartilages of visceral arches III through VI are cartilaginous **gill rays** that support the gill lamellae. The last pair of visceral cartilages (VII) does not possess a gill surface, but the hyoid arch (II) bears gill rays from the hyomandibular and ceratohyal cartilages and supports a single demibranch in the anterior wall of gill slit I.

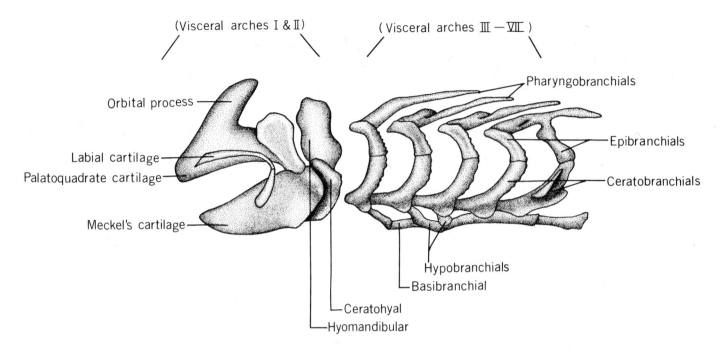

(Visceral arches I & II)   (Visceral arches III — VII)

Orbital process

Labial cartilage

Palatoquadrate cartilage

Meckel's cartilage

Pharyngobranchials

Epibranchials

Ceratobranchials

Hypobranchials

Basibranchial

Ceratohyal

Hyomandibular

Fig. 4-8. Lateral view of the visceral skeleton of *Squalus*.

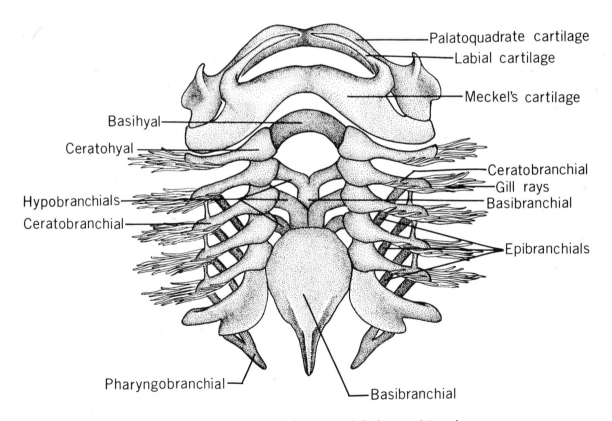

Palatoquadrate cartilage

Labial cartilage

Meckel's cartilage

Basihyal

Ceratohyal

Hypobranchials

Ceratobranchial

Ceratobranchial

Gill rays

Basibranchial

Epibranchials

Pharyngobranchial

Basibranchial

Fig. 4-9. Ventral view of the visceral skeleton of *Squalus*.

BONY FISHES

Within the bony fishes there is a general trend toward reduction in the number of posterior arches, and partial or total replacement of the visceral cartilages by bone. This bone may be in the form of dermal bone that replaces an area previously occupied by a visceral cartilage, or endochondral bone that replaces a given visceral cartilage.

The palatoquadrate cartilages contribute to the upper jaw only posteriorly where the **quadrate bone** replaces its posterior margin. Several pterygoid bones may replace the other parts of the palatoquadrate where it becomes associated with the braincase. An **articular bone** replaces the posterior margin of Meckel's cartilage, thus forming an articular-quadrate jaw suspensorium, an arrangement that persists throughout the amphibians, reptiles, and birds, to change only with the advent of the mammals. The anterior portion of Meckel's cartilage becomes ensheathed by several dermal bones (primarily the dentary), and if the cartilage remains at all, it may persist as a core inside the mandible.

As many as six endochondral bones may replace the hyoid cartilages, including two dorsal components, the **hyomandibular** and the **symplectic,** which may abut the quadrate to provide a hyostylic jaw mechanism. The remaining visceral arches resemble those of *Squalus* except that the gill cartilages are replaced by endochondral bone.

NEOTENIC AMPHIBIANS

In larval amphibians and in the tadpoles of anurans the visceral skeleton is similar to that encountered in the bony fishes; however, the number of gills is reduced, and those that remain are usually external. Study the visceral skeleton of *Necturus* following the diagram in **Fig. 4-10** and note the reduction in elements. The visceral skeleton consists of the cartilages of four visceral arches; the hyoid arch and the three posterior branchial arches. Though only four identifiable arches remain, portions of the posterior arches contribute to the laryngotracheal area.

TETRAPODS

Upon metamorphosis, the visceral skeleton becomes transformed into a structure adapted for a completely different mode of life, inspira-

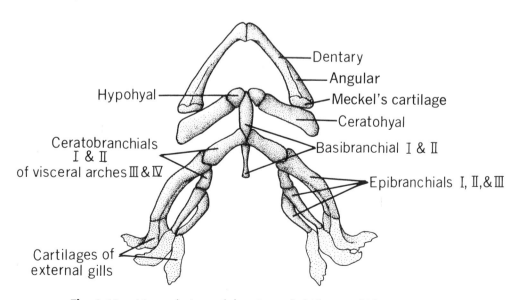

Fig. 4-10. Ventral view of the visceral skeleton of *Necturus*.

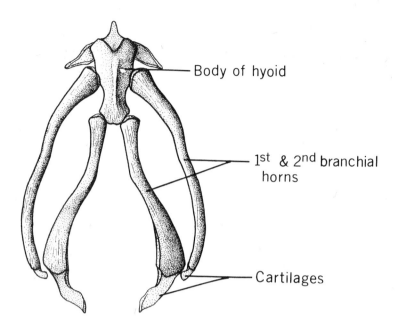

Fig. 4-11. Ventral view of the hyoid apparatus of a turtle *(Chelydra)*.

tion of aerial oxygen into the lungs. Thus, the anterior visceral arches of the tetrapods, instead of serving for gill support, are transformed into the jaws and structures for support of the tongue (**glossal skeleton** or **hyoid apparatus, Fig. 4-11**). In most reptiles the anterior horns and much of the body of the hyoid apparatus are derived from the hyoid arch, and the posterior horns from visceral arches III and IV. Posterior arches also participate in the formation of the opening to the trachea, glottis, and epiglottis, and in the formation of the laryngeal and tracheal rings. In addition, portions of the hyoid arch may participate in the formation of the middle ear, particularly one of the bony ear ossicles. The dorsal segment of the hyoid arch, the hyomandibular, becomes transformed into the **columella** or **stapes,** the single ear ossicle of amphibians, reptiles, and birds (see Fig. 4-20). The stapes serves to transmit sound waves from the tympanic membrane to the fenestra ovalis, where the aerial sound waves are transformed into fluid waves in the perilymph of the inner ear. Because the hyomandibular primitively articulated with the otic region, it was perfectly preadapted to assume a new role in the transmission of aerial sound waves. The

remainder of the hyoid arch contributes, along with portions of the posterior gill arches, to the formation of the hyoid apparatus.

It should be noted here that the entire middle ear complex is greatly reduced in urodeles. However, in most amphibians the middle ear cavity, which is derived from the first visceral pouch (as one might suspect from its location), is bounded on the exterior by the tympanic membrane, but communicates with the pharynx by the **eustachian tube,** a system that permits equalization of pressure. This system persists through all the tetrapods. The eustachian tubes will be observed in dissections of the cat.

MAMMALS

In mammals there is a well-developed hyoid apparatus, but it is not as large as that of other tetrapods; it is derived predominantly from the hyoid arch, with minor contributions from visceral arch III. Posterior arches contribute to the cartilages of the laryngeal and tracheal regions.

Examine the hyoid apparatus of the cat (**Fig. 4-12**). It is composed of anterior and posterior horns joined at the **basihyal bar.** The **anterior**

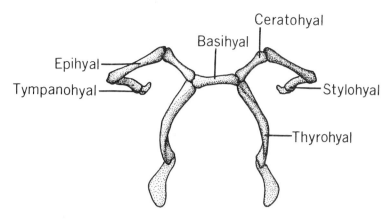

Fig. 4-12. Ventral view of the hyoid apparatus of the cat.

horn or **cornu** consists of **tympano-, stylo-, epi-,** and **ceratohyal** segments; these are all derivatives of the second visceral arch. The body of the hyoid is derived from arches II and III, and the posterior element, the **thyrohyal**, is misnamed, since it is a first-arch derivative.

Examine dried specimens of whole and cross sections of the larynx and anterior tracheal regions, and following **Fig. 4-13,** study the structures outlined below.

The esophagus, which is the anterior muscular tube of the alimentary canal, lies dorsal to the **trachea** and **larynx**. The trachea is the long cartilaginous tube that extends posteriorly from the larynx, and bifurcates to form the **bronchi** leading into the two lungs. The C-shaped tracheal cartilages prevent collapse of the trachea. The larynx is the chamber anterior to the trachea that houses the vocal cords. A cartilaginous **epiglottis** guards the entrance to the larynx and prevents the entry of food. The **glottis** is the actual slitlike opening to the larynx. The **thyroid cartilage** is the large unpaired cartilage that forms most of the ventral and lateral regions of the larynx; the **cricoid cartilage** is a smaller cartilage located posterior to the thyroid. A pair of very small **arytenoid cartilages** are located middorsally on the cranial border of the cricoid cartilage. **False vocal cords** extend from the base of the epiglottis to the arytenoid cartilages, and the **true vocal cords**, larger, lateral folds within the larynx, ex-

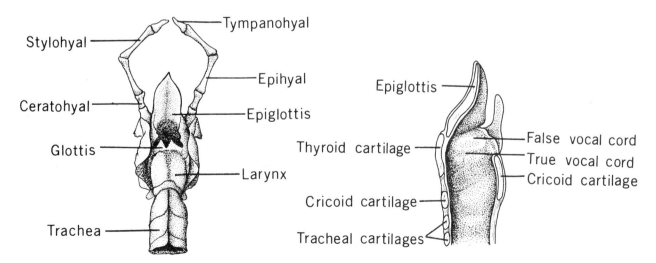

Fig. 4-13. Dorsal view and sagittal section of the larynx and trachea of the cat.

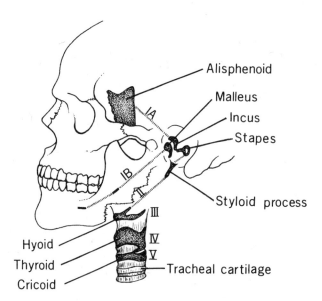

Fig. 4-14. Components of the visceral skeleton of man. IA, line connects palatoquadrate derivatives (alisphenoid and incus); IB, line connects vestiges and derivatives of Meckel's cartilage (remnants of cartilage in mandible and malleus); II, line connects hyoid arch derivatives (stapes, styloid process, and hyoid apparatus); III to V, third, fourth, and fifth arch derivatives. III, greater horn of hyoid. (Slightly modified after Kent, *Comparative anatomy of the vertebrates*, 3rd ed. Courtesy The C. V. Mosby Company, St. Louis.)

tend from the arytenoids to the thyroid cartilage. The vocal cords are set in motion by air entering the larynx, and are controlled to a degree by the movement of the arytenoid cartilages.

The cartilages or replacing bones that form the laryngeal and tracheal skeleton in most tetrapods probably represent the visceral skeleton of the last two visceral arches. Problems are often encountered in attempting to homologize these cartilages with specific visceral arches, but it is now thought that in mammals the cricoids and arytenoids are products of arches VI and VII, and the thyroids, arches IV and V. Anterior tracheal rings may be derived from the seventh arch, though this is uncertain. The cartilage of the epiglottis is not visceral skeleton.

Fig. 4-14 illustrates the contributions of the visceral skeleton to the head region of man.

Recall that in the early evolution of tetrapods the hyomandibular lost its function as a jaw suspension mechanism, and became transformed into the columella or stapes, the single ossicle of the middle ear of amphibians, reptiles, and birds. The middle ear cavity and stapes are best observed in the turtle skull (Fig. 4-20). Note also the position of the articular and quadrate bones in relation to the ear cavity (**Fig. 4-15A**).

As mammals evolved, there were changes in the jaw musculature, and they developed a much stronger jaw. One of the osteological changes that occurred was an increase in the size and importance of the dentary bone, with a concomitant diminution in the importance of the posterior bones of the lower jaw. Eventually in the therapsids a double jaw articulation developed, and an articulation between the dentary and the squamosal bones took over as the sole jaw suspensorium, leaving the small articular and quadrate bones free for other functions. The tympanic membrane was probably located directly posterior to the jaw joint in the therapsids, and it has been suggested that the small articular and quadrate bones may have

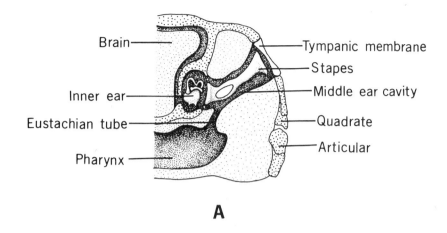

A

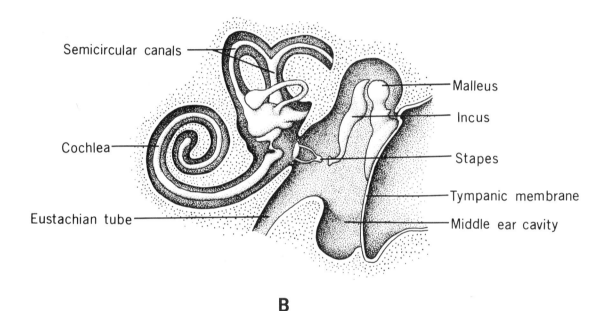

B

Fig. 4-15. Diagrammatic sections through the otic region of *A*, an amphibian, and *B*, a mammal. (After Romer, 1970.)

served even then to aid in the transmission of sound waves, perhaps from the lower jaw to the middle ear. However speculative this may be, in the first mammals one sees three fully developed ear ossicles (Fig. 4-15). Innermost is the **stapes,** inherited as the single ear ossicle from the reptiles. Then there are the **incus** and **malleus,** new names for the quadrate and articular bones, respectively, which migrated into the middle ear cavity to form two of the three mammalian ossicles, structures which define the class Mammalia most properly.

The Dermatocranium

The ancient fishes were heavily armored with thick, multi-layered dermal armor that covered most of the body, and developed within the dermis of the skin. In the cartilaginous fishes the armor was lost, but in the later fish and tetrapod lines the dermal armor became confined to the head and pectoral regions, and developed deep within the dermis of the skin. These dermal bones constitute the dermatocranium, and are the bones that superficially encase the structures of the head region.

FISHES

In the lines of bony fishes leading to tetrapods the dermatocranium was an extensive covering of numerous dermal bones. Though not on the main stream of tetrapod evolution, the Nile bichir (*Polypterus:* Chondrostei), or the bowfin (*Amia:* Holostei), will serve to illustrate certain features of the primitive condition. Study the head region of a bichir or bowfin following the description below and **Fig. 4-16.**

First examine the dorsal aspect of the skull, noting the roofing bones that cover the brain and neurocranium and contribute to the lateral walls of the skull. Note the position of the **quadrate;** it is the only bone to replace part of the palatoquadrate cartilage. Anteriorly, dermal bones form the margins of the jaw, and are lined by numerous **homodont** (uniform) **teeth.** A series of dermal bones forms the lower jaw. The **articular** bone forms the lower jaw articulation; it is a replacing bone formed in the posterior extent of Meckel's cartilage. Posteriorly there is an **opercular** and **gular** series of dermal bones. Off the porterior margin are the portions of the pectoral girdle that are also dermal in origin, being derivatives of the posterior dermatocranium.

In addition, the roof of the mouth is formed by a series of dermal bones, and in the midventral line an unpaired **parasphenoid** bone is present.

PRIMITIVE TETRAPODS

Although the skull of *Necturus* is somewhat aberrant, and is composed of a considerable amount of cartilage, it nonetheless shows some of the primitive features of the vertebrate dermatocranium. The skulls of both *Necturus* and a snapping turtle, *Chelydra,* will be considered here to gain some insight into the primitive tetrapod dermatocranium.

First, however, review the bones contributed by the neurocranium. These include the entire occipital series. In the specimens of *Necturus* (**Figs. 4-17** and **4-18**), locate the paired **exoccipitals** lateral and ventral to the **foramen magnum;** each exoccipital bears an **occipital condyle,** unlike the primitive tetrapods, which

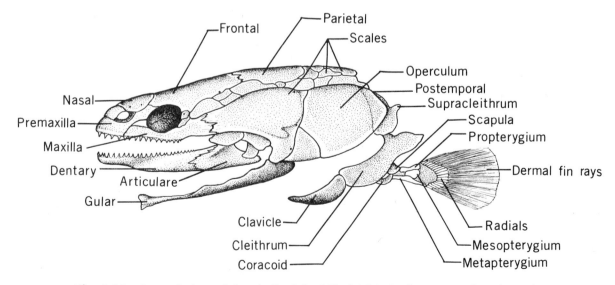

Fig. 4-16. Lateral view of the skull of the Nile bichir *(Polypterus:* Chondrostei) showing the dermal bones and the pectoral girdle.

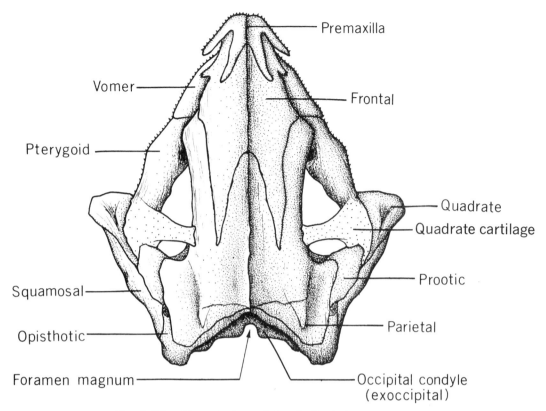

Fig. 4-17. Dorsal view of the skull of *Necturus*.

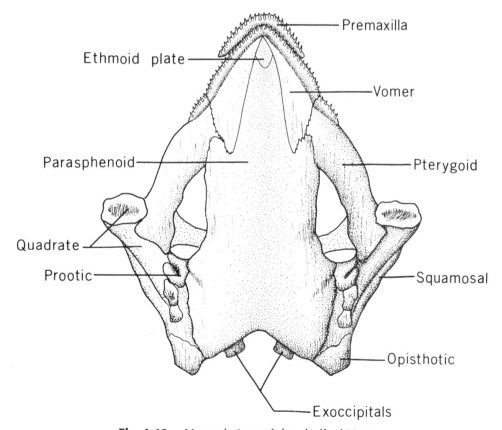

Fig. 4-18. Ventral view of the skull of *Necturus*.

had only one. In addition, the **prootics** and **opisthotics** of the otic region are neurocranial contributions. Also, a small part of the **ethmoid plate** of the neurocranium may be observed in the ventral aspect of the skull. The **quadrate** and **articular** bones are replacement bones of the posterior bones of the palatoquadrate and Meckel's cartilages.

In the turtle skull (**Figs. 4-19, 4-20,** and **4-21**), a similar series of neurocranial contributions may be observed. Locate these: they include the occipital series, the **supra-, ex-,** and **basioccipitals;** the **otic** bones, **pro-** and **opisthotic;** and the **basisphenoid,** plus other minor sphenoidal contributions. As in *Necturus,* the quadrate and articular are present as derivatives of the first visceral arch.

Returning to the skull of *Necturus,* compare it with Fig. 4-17 and 4-18. Though highly spe-

cialized (retaining few truly primitive features), the dermatocranium still serves to illustrate some basic features of the tetrapod skull. Ignoring neurocranial contributions, dorsally the roof is composed primarily of the large **parietal** and the **frontal** bones. The **squamosal** is a sliver of bone on the posterolateral margin. The V-shaped **premaxilla** is the anteriormost tooth-bearing element. Also forming the jaw margin are the toothed **vomers** that articulate with the **parasphenoid** and **pterygoids.** The most posterior marginal bones are the pterygoids, which bear teeth on their anterior margins. Most of the roof of the mouth is composed of the single parasphenoid.

Now examine the skull of *Chelydra,* and compare it with Figs. 4-19, 4-20, and 4-21. Ignoring neurocranial contributions, the dorsal roof is composed posteriorly of large **parietals.** Anterior

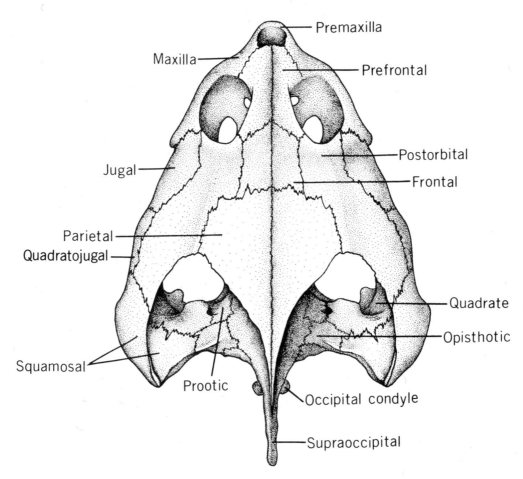

Fig. 4-19.   Dorsal view of the skull of *Chelydra.*

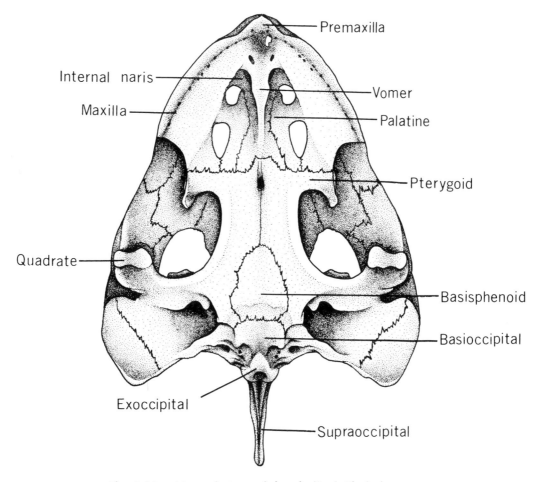

Fig. 4-20. Ventral view of the skull of *Chelydra*.

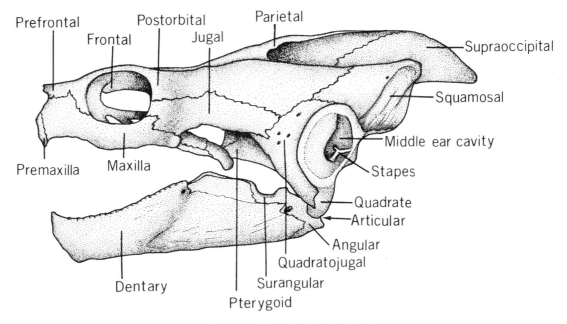

Fig. 4-21. Lateral view of the skull of *Chelydra*.

to the parietals is a smaller pair of **frontals,** and forming the medial walls of the orbits, a pair of **prefrontals** immediately posterior to the **external nares.** The nares, which are originally paired, form a single external opening in *Chelydra.* Note that the **internal nares** (**choanae**) located anteriorly in the palate are distinctive openings. Posterior to the orbits, the large **postorbitals** contribute to the dermal skull roof; they articulate posteriorly with the **squamosals,** which form the posterodorsal margin of the middle ear cavity. The middle ear cavity is bounded anteriorly by the **quadrotojugal,** which articulates anteriorly with the **jugal** bone. Within the middle ear cavity (tympanic cavity), the rodlike **stapes** may be seen projecting outward through a small hole in the dorsal area of the **quadrate.** Recall that the quadrate is the posterior part of the palatoquadrate. Medially the enlarged footplate of the stapes inserts into the **oval window** (**fenestra ovalis**) to come into contact with the inner ear.

The anterior margin of the upper jaw is formed by the **premaxilla** and **maxilla,** neither of which bears teeth in turtles. Now turn to the palate, which is composed of dermal bones. Anteriorly the primary bony palate is composed of the palatal processes of the premaxilla and maxilla. There is a median **vomer** (fusion of originally paired elements), and lateral to the vomer are located the **palatines.** Posteriorly the palate is formed by the large **pterygoids.**

Because of the structure of the primary palate, air cannot be taken into the lungs from the internal nares while feeding occurs. In some turtles and in all the crocodilians the premaxilla, maxilla, palatines, and pterygoids have become infolded to form a **secondary palate,** thus forming a chamber through which air may pass posteriorly in the skull while feeding occurs.

As in all amphibians, reptiles, and birds, the lower jaw is composed of multiple elements. The major dermal contribution is the large **dentary** bone. The posterolateral margin also contains the **surangular** and **angular** bones, in addition to other medial elements. The articular is an endochondral contribution of the posterior end of Meckel's cartilage.

MAMMALS

The general morphological features of the mammalian skull will be illustrated using the cat skull; however, a comparative study should be undertaken by observing many different mammalian skulls, including that of man (**Fig. 4-27**).

First consider the neurocranial contributions to the skull (**Figs. 4-22** and **4-23**). These features are observed in lateral and ventral views in the cat skull (**Figs. 4-24** and **4-25**), and in sagittal section (**Fig. 4-26**). Posteriorly the occipital region is formed by several occipital ossification centers, but in the adult fusion may occur to the extent that the individual elements are indistinguishable. Note that the **nuchal crest** is borne on the **occipital,** and the lateral **paroccipital processes** are extensions of the occipital. The **supraoccipital** forms the dorsal roof of the **foramen magnum** and the **exoccipitals** form its lateral borders, and give rise to the two lateral **occipital condyles,** typical of mammals. Study the region at which the junction of the **tympanic bullae** with the ex- and basioccipitals occurs; there is a large foramen known as the **jugular** or **posterior lacerate foramen** through which the cranial nerves IX, X, and XI are transmitted in company with the internal jugular vein. Posterior and medial to the posterior lacerate foramen is the small **hypoglossal canal,** which transmits cranial nerve XII. The **basioccipital** is a ventromedial unpaired bone that extends from the foramen magnum to the **basisphenoid.** The medial basi- and **presphenoid** bones are also contributions of the neurocranium. Note that in the sagittal section the primary area occupied by the brain is the

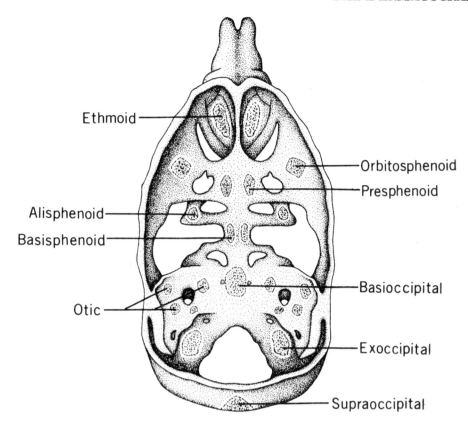

Fig. 4-22. Dorsal view of the neurocranium of a fetal pig showing the ossification centers typical of most mammals. (After Kent, *Comparative anatomy of the vertebrates,* 3rd ed. Courtesy The C. V. Mosby Company, St. Louis.)

As, Pleurosphenoid (alisphenoid)
Ba, Basioccipital
Bs, Basisphenoid
C, Cribriform plate of ethmoid
E, Ethmoid, perpendicular plate
Ex, Exoccipital
F, Frontal
I, Interparietal
N, Nasal
Os, Orbitosphenoid
Ot, Otic (petrous)
P, Parietal
Pa, Palatine
Pm, Premaxilla
Ps, Presphenoid
Pt, Pterygoid
So, Supraoccipital
Sq, Squamosal
Vo, Vomer

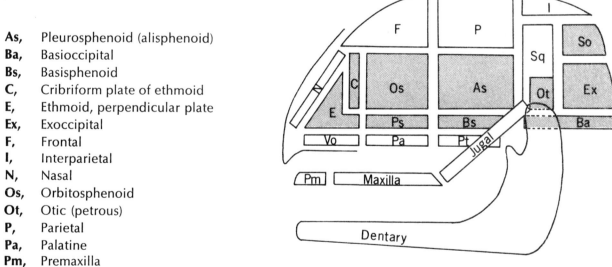

Fig. 4-23. Diagrammatic illustration showing the major components of the neurocranium (dark) and dermatocranium (white) in the mammalian skull. A key to the symbols used appears left. (Slightly modified after Kent, *Comparative anatomy of the vertebrates,* 3rd. ed. Courtesy The C. V. Mosby Company, St. Louis.)

middle cranial fossa, which houses the cerebral hemispheres. The saddle-shaped **sella turcica** is located in the floor of the cerebral fossa. It houses the hypophysis. The floor of the anterior region of the cerebral fossa is formed by the presphenoid. Note the large **sphenoidal air sinus** in the presphenoid.

In the posterolateral wall of the orbit, forming part of the sides of the braincase, is the winglike **alisphenoid** bone: it is a homologue of the reptilian epipterygoid, and a derivative of the mandibular arch. The alisphenoid projects into the ventral aspect of the skull as a pterygoid process that terminates posteriorly as a projection known as the **hamulus.** Study the three foramina that pierce the alisphenoid bone; these are best observed in the disarticulated bone. The anteriormost and largest is the

anterior lacerate foramen (orbital fissure), which transmits cranial nerves III, IV, and VI (leading to the extrinsic eyeball muscles), and also a branch (the ophthalmic) of cranial nerve V. The middle and smallest foramen, **the foramen rotundum,** transmits the maxillary branch of cranial nerve V, and the posterior foramen, the **foramen ovale,** transmits the mandibular branch of V.

Anterior to the alisphenoids are the median presphenoid and a pair of lateral **orbitosphenoids;** they are pierced by the large **optic foramen** through which the optic nerve (cranial nerve II) is transmitted.

Now turn to the sagittal section of the skull and note that the anterior wall of the anterior cranial fossa, which is the area for the olfactory bulbs, is formed by a plate of bone known as

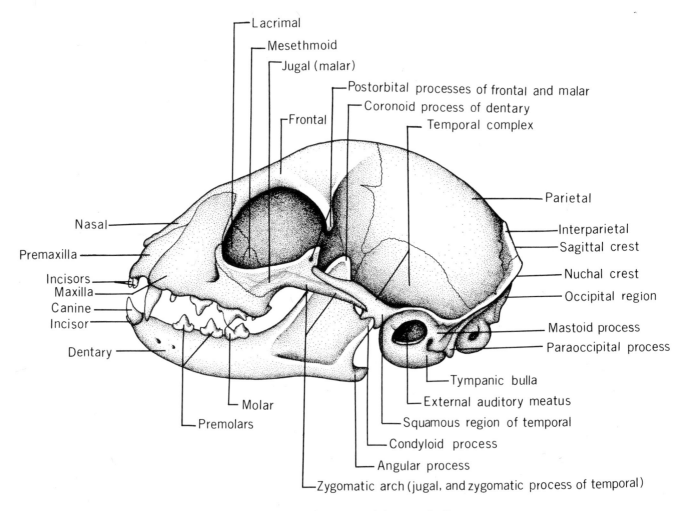

Fig. 4-24.   Lateral view of the cat skull.

the **cribriform plate;** it is derived from the mesethmoid bone. There is a perpendicular plate of bone (also of mesethmoid origin) that extends anteriorly to separate the two nasal cavities. The cribriform plate is pierced by many small foramina that transmit the small branches of the olfactory nerve (cranial nerve I). Each of the nasal cavities is filled with scrolls of thin bone known as the **turbinals** or **conchae;** they are covered by the nasal epithelium and serve to increase the surface area for olfaction.

Utilizing the sagittal sections and the disarticulated specimens, study the neurocranial contributions to the otic region. There are several centers of ossification in the region of the otic capsules that unite to form the **periotic** bone (**petrosal** or **petromastoid**). In the sagittal section one can observe the internal portion of the periotic, containing the inner ear; it lies in the ventral portion of the posterior cranial fossa and is pierced by the **internal auditory meatus.** The internal auditory meatus transmits cranial nerves VII and VIII. The eighth (auditory) nerve is a purely sensory nerve that emerges from the inner ear; the seventh (facial) emerges from the **stylomastoid foramen** beneath the tip of the mastoid process. There is a depression in the periotic dorsal to the meatus that houses a small portion of the cerebellum. Locate the **mastoid process** externally; this process plus the periotic region represent the primitive opisthotic and prootic elements.

The adult mammalian skull is not without contributions from the visceral skeleton, though they are somewhat reduced (see Fig. 4-14). Look within the external auditory meatus and try to find the small auditory ossicles within the middle ear cavity. Recall that the innermost

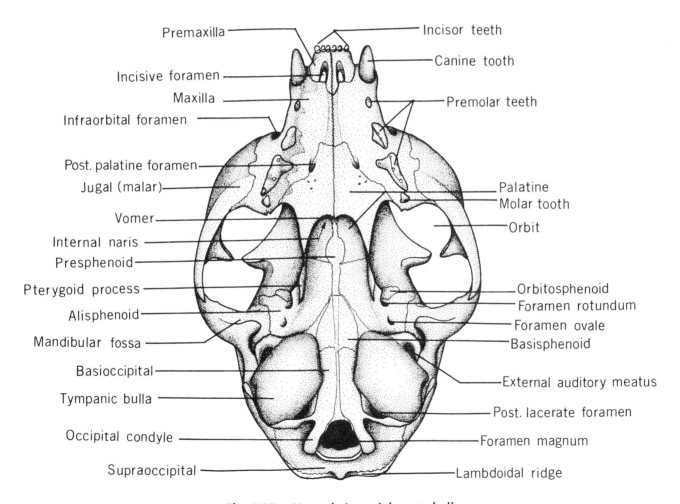

Fig. 4-25. Ventral view of the cat skull.

of these, the **stapes** (derived from the hyomandibular), was present in the amphibians, reptiles, and birds. Mammals have acquired an additional two: the **incus,** is derived from an endochondral replacement of the posterior portion of the palatoquadrate, and which formed the quadrate in the reptilian skull; and the **malleus,** derived from the posterior portion of Meckel's cartilage that formed the articular of reptiles. Note that the sequence of ossicles in the middle ear responds to the necessary sequence of evolutionary migration. The stapes has an oval-shaped footplate that fits into the oval window (fenestra ovalis) of the inner ear; where they touch, sound vibrations are transformed into fluid waves in the perilymph of the inner ear. The mammalian ear ossicles should be on demonstration for your examination. Examine a specimen in which the tympanic bulla

has been removed and study the openings in the ventral periotic. The more dorsal opening is the **oval window (fenestra ovalis)** into which the stapes fits (see the similar structure in an avian species, Fig. 4-29); the more ventral, the **round window (fenestra rotunda).** The entire mandible is invested by a single dermal bone, the **dentary.** Study the dentary bone, and note its posterior **angular process** and the **condyloid process** that articulates dorsally with the **mandibular fossa** of the squamous portion of the temporal bone, forming a dentary-squamosal jaw suspensorium characteristic of the class Mammalia. Note the differentiated tooth row on the dentary; from anterior to posterior, there are **incisors, canines, premolars,** and **molars.** The lateral surface of the dentary is occupied by the large **coronoid fossa,** which serves for the insertion of jaw muscles. Note the **coronoid proc-**

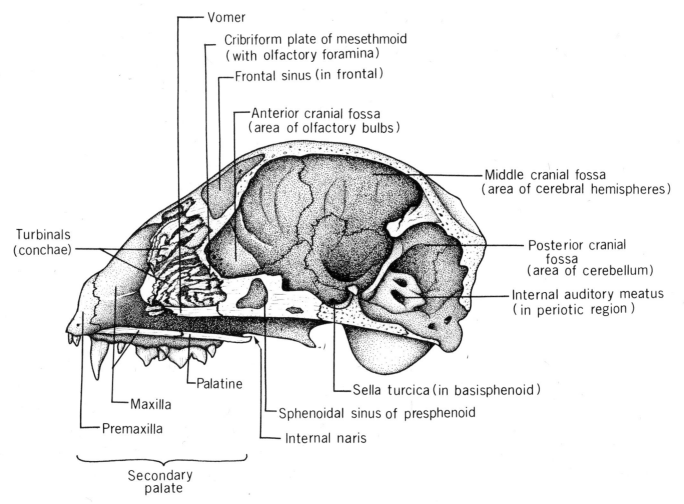

Fig. 4-26. Sagittal section of the cat skull.

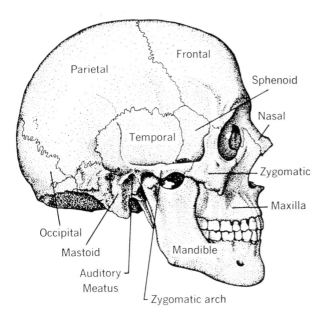

Fig. 4-27. Lateral view of human skull. (From Torrey, *Morphogenesis of the vertebrates,* 3rd ed., Fig. 12-22. Courtesy John Wiley & Sons, Inc., New York.)

ess, to which the temporal muscles attach, and the **angular process,** to which some mandibular muscles attach. On the medial surface of the dentary are located the large **mandibular** and smaller **mental foramina,** which transmit a branch of the mandibular portion of cranial nerve V (the trigeminal), which innervates the teeth and portions of the lower jaw. The nerve enters through the mandibular foramen and exits through the mental foramen.

The actual mammalian dermatocranium is composed of dermal bones that superficially surround most of the brain and associated sense organs dorsally and laterally. Examine both articulated and disarticulated skull elements in studying the cat dermal skull.

Forming the ventral and lateral borders of the **external nares** are the **premaxillae,** which contain the incisor teeth and are pierced posteriorly by the **incisive foramen** (anterior palatine foramen). The latter transmits a portion of the maxillary branch of cranial nerve V (trigeminal) to the roof of the mouth. Posterior to the premaxillae are the large **maxillae,** which bear the canines, premolars, and molars. The

maxillae complete the upper jaw, and form, along with the premaxillae and **palatines,** the **secondary palate.** They contribute posteriorly to the anterior region of the **zygomatic arch.** The maxillae are pierced laterally by the **infraorbital foramen,** which also transmits a branch of the maxillary division of cranial nerve V. The palatines form the posterior region of the secondary palate. They are pierced by the **posterior palatine foramen,** which also transmits a branch of the maxillary division of V to the mouth. The **internal nares** emerge posteriorly above the palatines. There is a single median **vomer** (originally paired) anterior to the **presphenoid** in the primary palate.

The anterolateral wall of the orbit is formed by the small and delicate **lacrimal** bone, which is pierced by the **nasolacrimal canal** for the tear duct. A very small bone posterior to the lacrimal in the medial wall of the orbit is the **mesethmoid;** it is a neurocranial contribution of endochondral bone.

The large zygomatic arch is formed by the **jugal (malar)** bone ventral to the orbit, and an anterior extension of the **temporal** bone. The **temporal fossa** (housing jaw muscles) is the large fossa posterior to the orbit. The temporal bone in mammals is a complex formed by the fusion of many elements; some have already been discussed in relation to the ear. The zygomatic portion and the section encasing the lateral portion of the brain are formed by the **squamosal** bone. The tympanic cavity is encased by a large **tympanic bulla,** which is formed from several separate dermal elements. In addition, there are the contributions from the visceral skeleton and neurocranium mentioned earlier.

The top of the skull is formed by several dermal bones. Anteriorly are the small paired **nasals;** the large **frontals,** which form most of the anterior roof of the brain, also contribute to most of the orbit. Note in the sagittal sections the **frontal sinus** and the extent of the frontals in forming the **middle cranial fossa.** The large paired **parietals** are present posteriorly. In

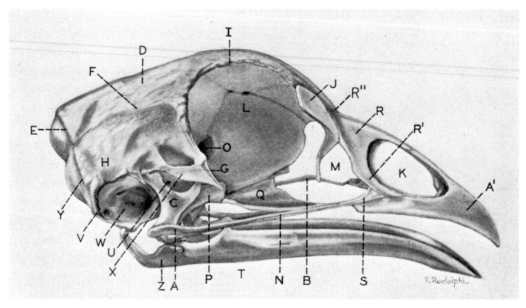

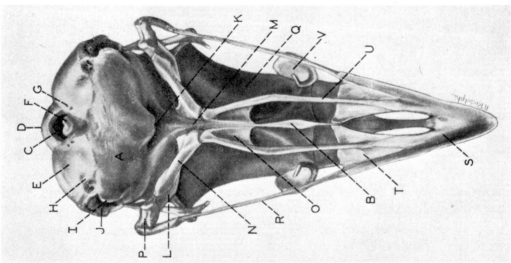

Fig. 4-28. Lateral (*above*) and ventral (*below*) view of the skull of a chicken. Lateral view: *A*, quadratojugal; *B*, vomer; *C*, quadrate; *D*, frontal; *E*, external occipital crest; *F*, lateral parietal crest; *G*, postorbital process; *H*, squamosal; *I*, supraorbital margin of frontal; *J*, prefrontal (lacrimal); *K*, external naris; *L*, interorbital septum; *M*, maxillary sinus; *N*, jugal bar (area of jugal bone); *O*, optic nerve foramen; *P*, pterygoid; *Q*, palatine; *R*, nasal; *S*, maxilla (zygomatic process); *T*, mandible; *U*, foramen lacerum oralis; *V*, opisthotic; *W*, external auditory meatus; *X*, zygomatic process; *Y*, margin of occipital region; *Z*, caudal extremity of mandible; *A'*, premaxilla; *R'*, lateral ramus of nasal; *R''*, nasal-frontal hinge.

Ventral view: *A*, basitemporal plate; *B*, vomer; *C*, foramen magnum; *D*, supraoccipital; *E*, exoccipital; *F*, occipital condyle; *G*, hypoglossal foramen; *H*, foramen lacerum aboralis; *I*, margin of basitemporal plate; *J*, external auditory meatus; *K*, opening for eustachian tube; *L*, quadrate (orbital process); *M*, parasphenoid rostrum; *N*, pterygoid; *O*, palatine; *P*, condyle of quadrate; *Q*, orbital fossa; *R*, jugal bar (area of jugal bone); *S*, premaxilla; *T*, maxilla; *U*, maxillary process of palatine; *V*, prefrontal (lacrimal). For the purposes of the laboratory exercise, only the following structures need be studied: Lateral view: *A–D, H, J–L, N, O, P–T, U–X, A', R''*. Ventral view: *A–F, J–P, R–T, V*. (From Chamberlain, 1943. *Atlas of avian anatomy.* Courtesy Michigan Agricultural Experiment Station.)

young individuals there may be a single **inter-parietal** in the middorsal line posterior to the parietals.

THE AVIAN SKULL

The avian skull (**Figs. 4-28** and **4-29**) is one of the most highly specialized among the vertebrates; many elements are fused, often leaving in the adult little trace of original sutures. But perhaps the greatest specialization is found in the **kinetic mechanism** of the skull, a mechanism that allows the upper jaw to move with respect to the rest of the braincase. The upper jaw swings upward and downward and articulates posteriorly with the braincase by means of a movable joint, the **nasal-frontal hinge** (Fig. 4-28). In addition, the quadrate is fully movable and is connected indirectly to the upper jaw via the palatal bones, the pterygoids and palatines, which slide along the parasphenoid rostrum. When the quadrates are rotated for-ward, the posteroventral aspects of the maxillae are pushed up and the upper jaw is thus pushed upward. When the quadrates are rotated backward, the palate is retracted and the upper jaw is thus depressed. This kinetic mechanism is operated by a very complex jaw musculature. Having a kinetic skull makes possible a fast-closing jaw mechanism and a wide gape, and also allows maintenance of the skull in its primary axis while jaw movement occurs.

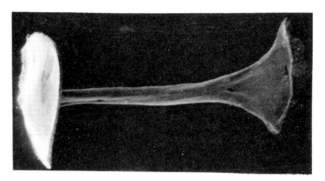

Fig. 4-29. Scanning electron micrograph of the bony stapes of a bird of paradise, illustrating the shaft and the footplate which fits into the oval window (fenestra ovalis). Actual length, 1.5 mm. (Photo by the author.)

5

The Postcranial Skeleton

The Postcranial Axial Skeleton

The postcranial axial skeleton consists of the notochord, vertebral column, ribs, and sternum. These features will be examined in the shark, the mud puppy, and the cat.

FISHES

In most fishes the vertebral column is not well differentiated into distinctive regions, and only trunk and caudal vertebrae are truly distinguishable. Examine special mounted preparations of the vertebral column of *Squalus* from both trunk and caudal regions, and study them following **Fig. 5-1.** First examine a cross section of a trunk vertebra, and note that it consists of a biconcave **centrum** that forms the body of the vertebra. Recall that the primitive axial skeletal element in the vertebrates is the **notochord.** In *Squalus* and in most fishes, the notochord is still present and extends down the canal through the center of the centrum. Biconcave vertebrae are termed **amphicoelous.** The dorsal arch of the centrum is the **neural arch;** it protects the spinal cord, which passes through its neural canal. In a sagittal section note that between each pair of neural arches is interposed another block, the triangular **intercalary plate.** The dorsal roots of spinal nerves emerge through the foramina in the intercalary plates; the ventral roots, through foramina in the neural arches. **Basapophyses** are present on the ventrolateral border of each trunk vertebra; the small rib cartilages are attached to these projections.

Study a series of caudal vertebrae and note that they differ from those of the trunk region in having ventral projections off the centrum, known as the **hemal arches,** which terminate as **hemal spines.** The **hemal canal** houses and protects the caudal vein and caudal artery.

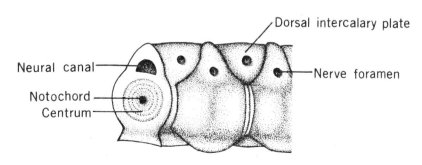

Fig. 5-1.   Trunk vertebrae of *Squalus*.

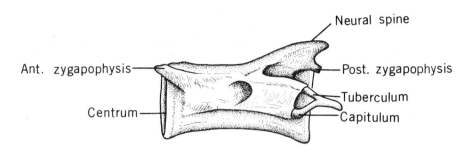

Fig. 5-2. Trunk vertebra of *Necturus*.

PRIMITIVE TETRAPODS

There is an enormous amount of variation in the vertebrae of tetrapods, and homologies are often difficult to trace. However, it is generally agreed that in the primitive tetrapod condition the vertebral centra were formed by three pairs of bony blocks surrounding the segmentally constricted notochord. Neural and hemal arches were present. This type of vertebra, termed **rhachitomous,** was typical of the labyrinthodont amphibians. Beyond this stage, tetrapod vertebrae have evolved in diverse patterns.

Examine the vertebral column of *Necturus* and distinguish the major regions. There is one **cervical vertebra;** those succeeding to the level of the pelvic girdle are **trunk vertebrae.** All of the trunk vertebrae possess free ribs. There is one **sacral vertebra;** it also has free ribs, but is modified for articulation with the girdle. The **caudal vertebrae** are those posterior to the pelvic girdle; they are distinguished by their possession of hemal arches.

Study the mounted and isolated trunk vertebrae (**Fig. 5-2**). Above the biconcave centrum is the **neural arch** through which runs the **neural canal.** A neural spine projects dorsocaudally from the posterior region. Anteriorly and posteriorly there are two sets of processes for articulation with successive vertebrae, the **anterior** and **posterior zygapophyses.** The anterior facets face dorsally; the posterior, ventrally. The zygapophyses, which serve to interlock the vertebral column, are an innovation with terrestrial vertebrates and are not present in fishes, where water supports the body.

The biconcave, amphicoelous centra contain remnants of the constricted notochord. Transverse processes project posterolaterally from the centrum; they form dorsal **diapophyses** and ventral **parapophyses** that articulate with the **tuberculum** and **capitulum** of the ribs, respectively. The distal rib is called the shaft. The sternum in *Necturus* is greatly reduced and is present only as a series of small cartilages.

MAMMALS

Study the mounted and disarticulated postcranial axial skeleton of the cat, following **Figs. 5-3, 5-4,** and **5-5.** The vertebral column is well differentiated into five major regions.

With few exceptions, there are seven **cervical vertebrae** in mammals, all of which lack free ribs. All but the last of the cervical vertebrae may be recognized by characteristic **transverse processes** pierced by the **transverse foramen,** which transmits the vertebral artery. The last cervical vertebra lacks the foramen and resembles a thoracic vertebra.

The first two cervical vertebrae are highly modified and deserve special attention. The first is called the **atlas;** it is a rounded vertebra with large transverse processes pierced by transverse foramina. The flattened neural arch lacks a neural spine but is perforated laterally by a pair of **atlantal foramina,** which transmit the vertebral arteries to the skull; the first of the spinal nerves exits the spinal cord through these foramina. The neural arch articulates with the two occipital condyles anteriorly; posteriorly,

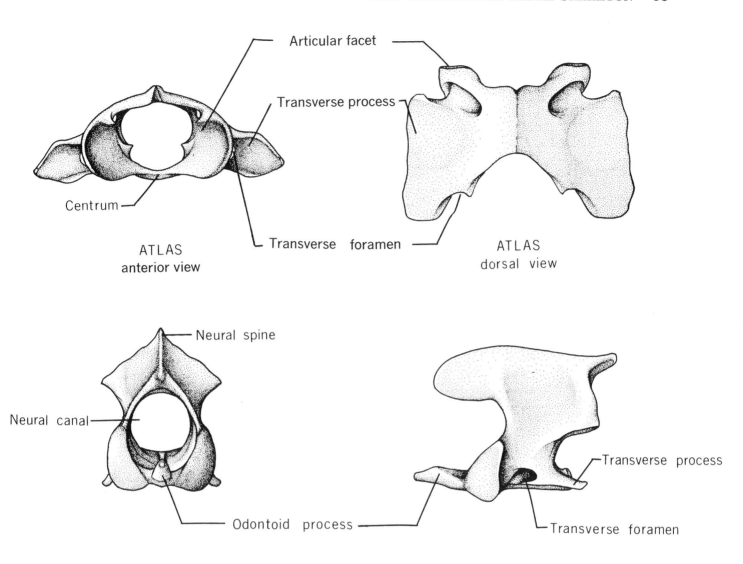

Articular facet

Transverse process

Centrum

ATLAS
anterior view

Transverse foramen

ATLAS
dorsal view

Neural spine

Neural canal

Odontoid process

AXIS
anterior view

Transverse process

Transverse foramen

AXIS
lateral view

Fig. 5-3.   The atlas and axis vertebrae of the cat.

it articulates with the centrum of the second cervical vertebra, the **axis.** The atlas was present in the first amphibians but the axis first appeared with the reptiles. Together they allow the free head movement characteristic of reptiles, birds, and mammals.

The axis is characterized by its long **odontoid process,** which projects into the atlas. Its neural arch is elongated and extends over the arch of the atlas.

There are thirteen **thoracic vertebrae;** all bear free ribs. Examine the characteristic neural spine. There are anterior and posterior zygapophyses. As in *Necturus,* the anterior facets face dorsally; the posterior, ventrally. The centrum has a flat surface on both ends; this condition is termed **acoelous.** Fibrous **intervertebral discs** are present in the living organism; these are very large in man. Transverse processes (diapophyses) extend laterally from the neural arches; each possesses an articular facet to accommodate the tuberculum of a rib. In some members of the thoracic series (anterior and middle portion) the capitulum of the ribs

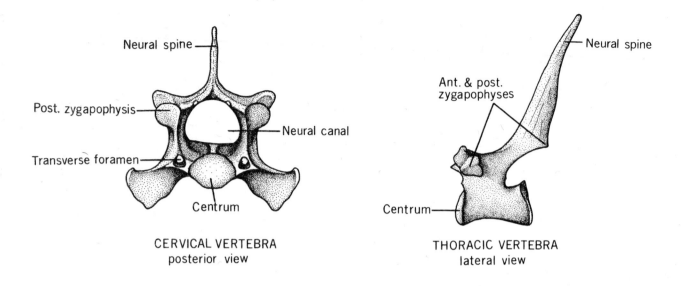

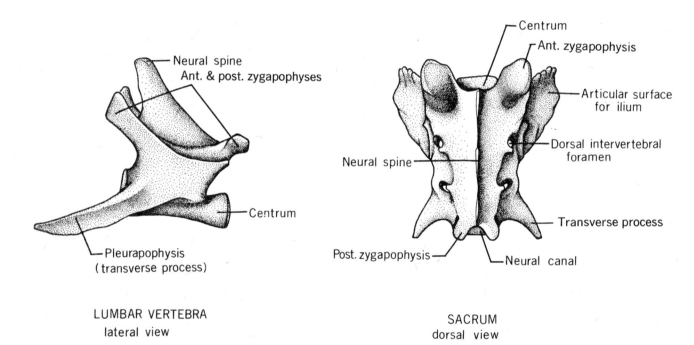

Fig. 5-4. Selected vertebrae of the cat.

may articulate between successive vertebrae. Note that each thoracic vertebra differs in details from the others.

Lumbar vertebrae are distinguished by their large size and their possession of long transverse processes (**pleurapophyses**); these represent a rib joined with a diapophysis. There are seven lumbar vertebrae.

The triangular **sacrum** represents the fusion of three individual vertebrae. Locate the individual neural spines and zygapophyses, and the posterior articular surface for the ilium.

The **caudal vertebrae** vary considerably in number, but are small compared to the others. The more anterior members of the series have the features characteristic of other vertebrae,

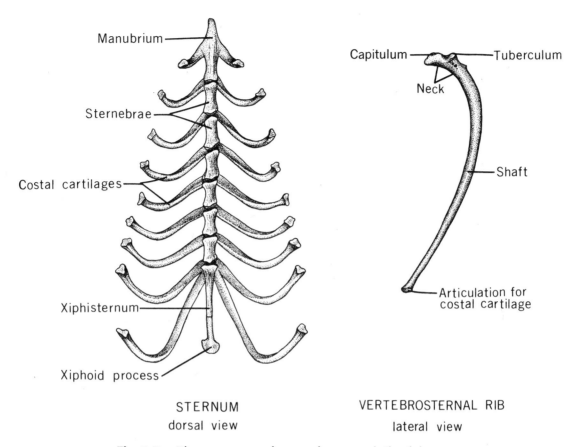

STERNUM
dorsal view

VERTEBROSTERNAL RIB
lateral view

Fig. 5-5. The sternum and a vertebrosternal rib of the cat.

but the more posterior lack these features. At about the level of the third caudal vertebra are small hemal processes ventrally. There may be small, V-shaped chevron bones articulating with these to form a hemal arch enclosing a hemal canal. The small chevron bones may be missing in the prepared mounts.

Examine articulated and disarticulated ribs. Most of the ribs have both a proximal head, the **capitulum**, and the more distal **tuberculum**; the area between the head and tubercle is the **neck. Costal cartilages** articulate distally with the shaft. Ribs that articulate with the sternum via the costal cartilages are termed **vertebrosternal ribs**; those not connected with the sternum are called **vertebral ribs**. Some have costal cartilages that attach with other costal cartilages; they are termed **vertebrocostal ribs**. The sternum is composed of endochondral

bony segments, **the sternebrae.** The first sternebrum is the **manubrium**, the last, the **xiphisternum.** There is a cartilaginous **xiphoid process** that extends caudally from the xiphisternum.

The Appendicular Skeleton

The appendicular skeleton is composed of the cartilages and bones of the pectoral (shoulder) and pelvic (hip) girdles and their respective fins or limbs. Although the appendicular skeleton is principally endoskeletal, there are some dermal contributions to the pectoral girdle; these are elements originally associated with the superficial dermal armor. In addition, heterotopic bones may develop in such diverse

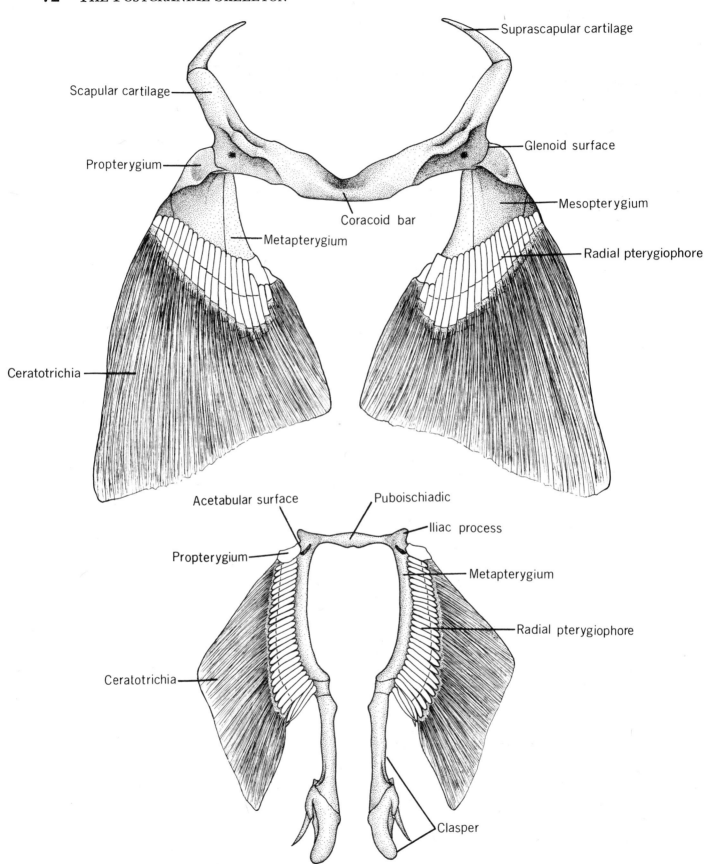

Fig. 5-6. Ventral view of the pectoral girdle *(above)* and pelvic girdle *(below)* of *Squalus*.

locations as the interventricular septum of ungulates and the penis of various other mammals. Sesamoid cartilages or bones develop in the tendons of the appendages of many vertebrates; the patella or kneecap is among the best known.

FISHES

With the exception of the chondrichthyian fishes, the **pectoral girdle** is a structure of dual origin, having both dermal and endochondral elements. The endochondral elements are primarily those forming the fin supports, while the dermal elements attach the girdle firmly to the body. Typically, the major contributions of the endoskeleton are the **coracoid, scapula,** and **suprascapula;** dermal contributions include (from ventral to dorsal) the **clavicle, cleithrum,** and **supracleithrum.** In the basic pattern a **posttemporal** unites the supracleithrum with the dermatocranium. The pectoral girdle of *Polypterus* (see Fig. 4-16) exhibits many features of the basic pattern.

Unlike the pectoral girdle, the **pelvic girdle** is entirely of endoskeletal origin. Typically,

there is a pair of cartilaginous or endochondral **puboischiadic** (pelvic) **plates** that unite in the midventral line at the **pubic symphysis.** In most tetrapods two major ossification centers form the **pubis** anteriorly, and the **ischium** posteriorly, with an **ilium** forming dorsally on each side.

Squalus · The appendicular skeleton of *Squalus* exhibits most of the features of the basic fish pattern but, of course, lacks the dermal contributions to the pectoral girdle.

Study the pectoral girdle and fin of the shark (**Fig. 5-6**) and note that the U-shaped girdle consists of a median ventral **coracoid bar** articulating laterally with the **scapular cartilages,** which have a **glenoid fossa** for articulation with the fin. A **suprascapular cartilage** may be distinguished at the top of the bar. The pectoral fin consists of three **basal pterygiophores**—from anterior to posterior, the **pro-, mesa-,** and **metapterygium.** Distally there is a series of **radial pterygiophores,** which meet the distal fibrous fin rays, called the **ceratotrichia.**

The pelvic girdle of *Squalus* is composed of a single **puboischiadic** cartilaginous bar that projects dorsolaterally as **iliac processes.** The **basal pterygiophores** consist of only two car-

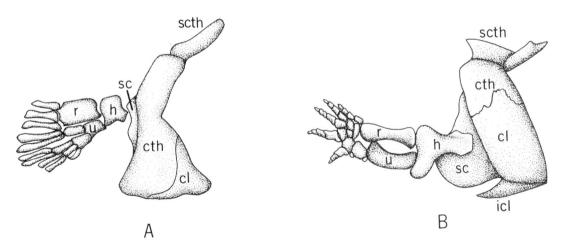

Fig. 5-7. Pectoral girdle and appendage of a crossopterygian (A), and a labyrinthodont (B). Abbreviations: cl, clavicle; cth, cleithrum; h, humerus; r, radius; sc, scapula; scth, supracleithrum; u, ulna. (Modified from Romer, 1970 and Gregory.)

tilages, the **propterygium** and the long **metapterygium.** As in the pectoral fin there is a series of **radial pterygiophores** distally on the ceratotrichia, which develop in the dermis. In the male *Squalus* there is a highly modified radial pterygiophore that forms the **clasper;** through it sperm are transmitted to the female cloaca.

PRIMITIVE TETRAPODS

In the evolution of tetrapods the crossopterygian fin is transformed into a limb designed for terrestrial locomotion (**Fig. 5-7**). In the first tetrapods (amphibians), and in the early reptiles, the limbs were almost at right angles to the body and served along with the fishlike

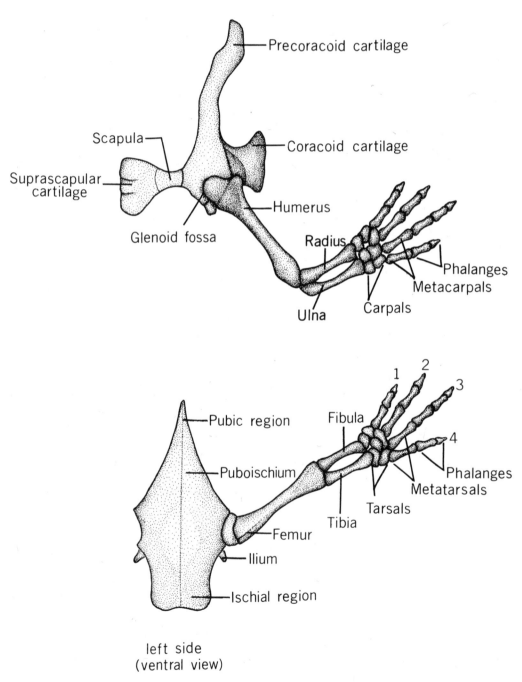

left side
(ventral view)

Fig. 5-8. Ventral view of the girdles and appendages of *Necturus*.

undulatory movements of the body for locomotion. However, in the more advanced reptiles the limbs became increasingly more important, and they were rotated beneath the body. Concomitant with changes in the limbs was a general strengthening in the pectoral and pelvic girdles.

Necturus · The appendicular skeleton of *Necturus* is aberrant in many features, but nonetheless illustrates the type of limb used for locomotion in the primitive sprawled posture. Examine a skeleton of *Necturus* (**Fig. 5-8**). The major

subdivisions of the limbs are the **brachium (humerus)**, **antebrachium (radius** and **ulna)**, and **manus (carpals, metacarpals,** and **phalanges)**. In the pelvic appendage the corresponding elements are the **thigh (femur)**, **shank (tibia** and **fibula)**, and **pes (tarsals, metatarsals,** and **phalanges)**. Note that only four toes are present.

The pectoral girdle consists of a **scapula** with its **suprascapular cartilage,** a **coracoid,** and a **precoracoid** extending anteriorly. There is a **glenoid fossa** for the articulation of the proximal end of the humerus.

The pelvic girdle is composed of a broad **puboischiadic plate** ventrally; anterior (**pubic**)

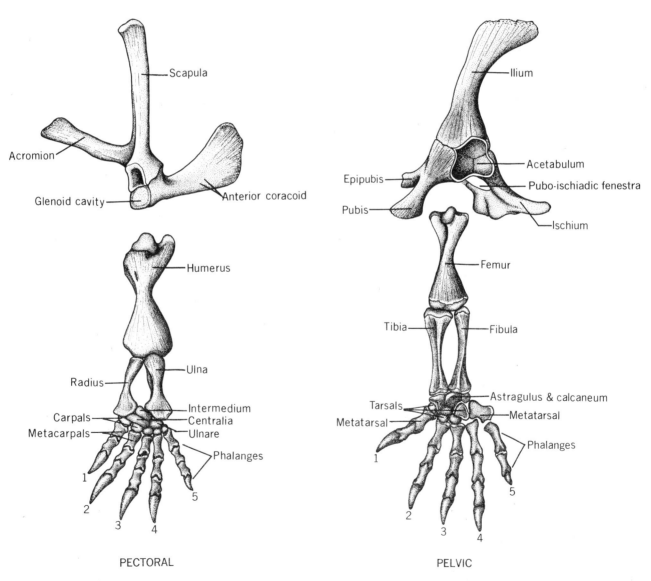

PECTORAL

PELVIC

Fig. 5-9. Lateral view of the left pectoral and pelvic appendages of *Chelydra.*

and posterior (**ischial**) regions may be distinguished. A small narrow **ilium** is present and extends dorsally from the **acetabulum** (articular fossa for the femur).

Chelydra · The appendicular skeleton of the turtle is specialized in many features, but serves to illustrate the reptilian type in which the brachium and thigh are held in the transverse plane of the body, with the antebrachium and shank at right angles.

Examine the appendicular skeleton of the turtle (**Fig. 5-9**), and note that in keeping with the evolutionary trend in tetrapods, the dermal portions of the pectoral girdle are greatly reduced. In turtles there are some dermal elements but these are incorporated into the **plastron** (ventral shell). The endoskeletal girdle consists of a **scapula**, with its **acromion process;** there is a posteroventral process, the **anterior coracoid.** The **glenoid cavity** articulates with the proximal end of the humerus. Study the prominent bones of the arm. They are the sturdy **humerus,** which articulates distally with the **radius** and **ulna.** The manus consists of a proximal row of **carpals** and five **metacarpals.** There are five **phalanges.** Note that the proximal row of carpals consists of **ulnare, intermedium,** and **centralia.**

Examine the pelvic appendage of the turtle (Fig. 5-9) and note that the **acetabulum** (for the articulation of the femur) is formed by three elements, the dorsal **ilium,** the anteroventral **pubis** (with its **epipubic process**), and posteroventral **ischium.** The ilium articulates with two sacral vertebrae. The pubes and ischia of opposite sides meet in the midventral line at a symphysis. A **pubo-ischiadic fenestra** is present between the pubis and ischium; it is the site of the origin of a muscle.

The sturdy bones of the thigh and shank are the **femur, tibia,** and **fibula.** The **pes** is composed of **tarsals, metatarsals,** and **phalanges.** The **astragulus** and **calcaneum** are fused as a single element; there are four distal tarsals.

MAMMALS

Examine the girdles and appendages of the cat (**Figs. 5-10, 5-11,** and **5-12**), using both articulated mounts and disarticulated bones.

The pectoral girdle is quite unlike those of the lower tetrapods. In keeping with the general trend in the evolution of the tetrapod pectoral girdle, the dermal portions of the girdle are greatly reduced, and the **scapula** is the primary element. The **glenoid fossa** for the articulation of the head of the humerus is directed ventrally instead of laterally as in the reptiles, and the limbs are directed ventrally beneath the body. The spine of the scapula separates the **supra-** from the **infraspinous fossae. The** coracoid is reduced to the small **coracoid process** on the anterior edge of the glenoid fossa. The ventral portion of the scapula spine continues as the **acromion,** which serves for the articulation with the **clavicle** in those mammals with a prominent clavicle. The clavicle of the cat is a very small bone embedded between the clavotrapezius and clavodeltoid muscles; it is the only portion of the dermal pectoral girdle to remain. The **metacromion** extends posteriorly from the ventral aspect of the **scapular spine.**

The **humerus** has a large rounded head that articulates with the **glenoid cavity.** There is a **greater tuberosity** at the proximal end. Distally there is an **entepicondylar foramen** for the passage of a nerve and blood vessel, and two large **trochleae** for articulation with the **radius** and **ulna.** The ulna articulates with the humerus by its **trochlear notch;** proximal to the notch is the **olecranon.** Distally the notch is bordered by the **coronoid process.** The **radial notch** accommodates the head of the radius. The **lateral styloid process** of the ulna articulates with the wrist. The radius has a disc-shaped head for rotation of the humerus and ulna. The **radial tuberosity** serves for the attachment of the biceps muscle. Distally the **medial styloid process** articulates with the wrist. The manus of the cat (Fig. 5-11) consists of the **carpus, meta-**

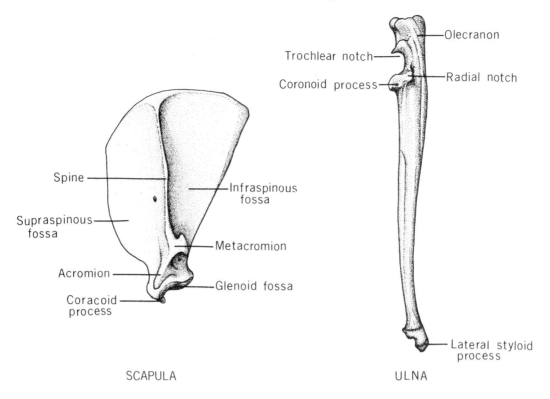

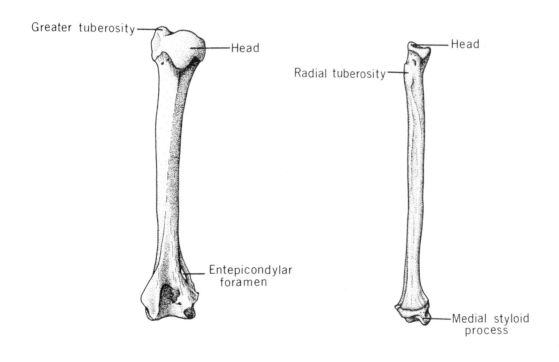

Fig. 5-10.   Pectoral bones of the cat (left side).

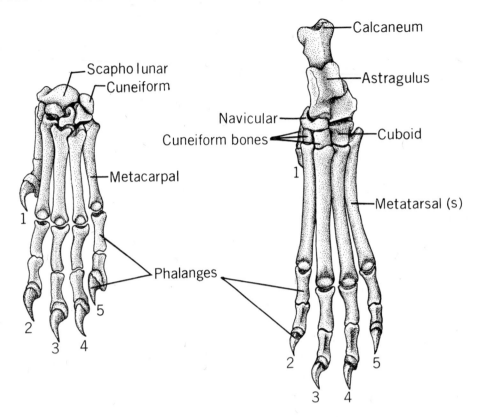

Fig. 5-11. Dorsal view of the left manus (left) and pes (right) of the cat.

carpus, and **phalanges.** The carpus contains two rows of small bones. The proximal row consists of, from medial to lateral, the **scapholunar** and the **cuneiform.** The small **pisiform** (frequently lost in preparation) articulates with the cuneiform. The large scapholunar represents the fused radiale, intermedium, and centrale. The pisiform is a sesamoid bone, serving for the attachment of muscle tendons. Many sesamoids develop all over the vertebrate body. The second row of carpals consists of four small bones. The five **metacarpals** form the palm of the hand, and the phalanges form the free fingers. Note that the medial digit is numbered 1.

The pelvic girdle of the cat (Fig. 5-12) is formed by two **innominate bones,** which represent the fused ilium, ischium, and pubis. These are visible as separate bones in the young animal. The **ilium** projects dorsally from the **acetabulum,** the fossa for articulation with the head of the femur. The large opening between the **ischium** and the **pubis** is called the **obtur-**ator foramen. The ischia and pubes unite ventromedially in a broad symphysis, forming with the sacrum a complete pelvic canal.

The **femur** exhibits a large head for articulation with the acetabulum. The **greater** and **lesser trochanters** are for muscle attachments. Distally there are two large condyles; the depression between them is called the **intercondyloid fossa.** The **epicondyles** are rough areas above the condyles. Anteriorly on the distal end is the fossa over which the **patella** (kneecap), a small sesamoid bone, glides. The shank consists of the **tibia** and **fibula.** The condyles on the distal end of the tibia are for the femur articulation; there is a large tuberosity that serves for the attachment of the patellar ligament. Distally is the **medial malleolus,** which articulates with the ankle. The slender **fibula** has a small head for articulation with the tibia; distally there is a process, the **lateral malleolus.**

The foot of the cat consists of the **tarsus, metatarsus,** and **phalanges.** The tarsus consist of seven bones. The long lateral heel bone is

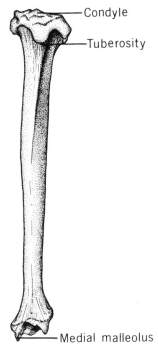

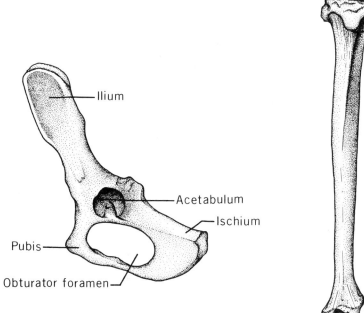

PELVIS

TIBIA

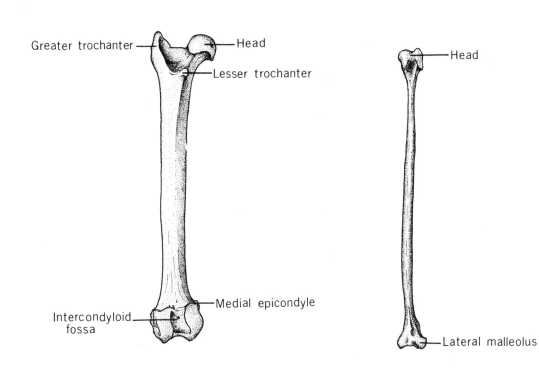

FEMUR

FIBULA

Fig. 5-12. Pelvic bones of the cat (left side).

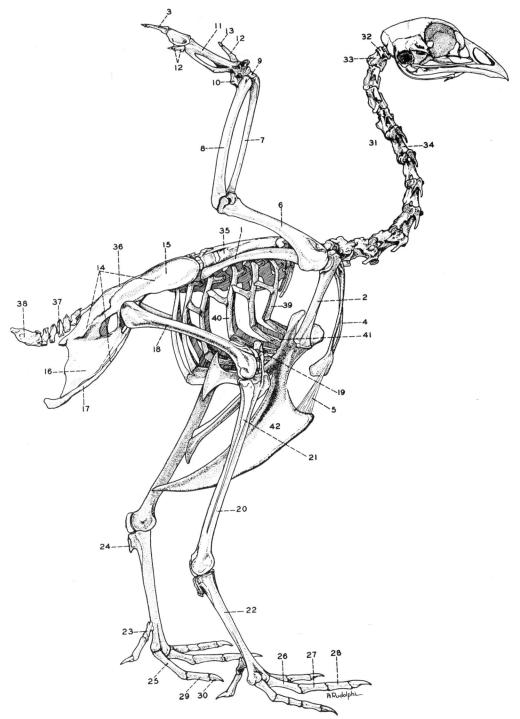

Fig. 5-13. Articulated chicken skeleton.

1, scapula; 2, coracoid; 3, second phalanx, third digit; 4, furculum (fused clavicles); 5, hypocledial ligament; 6, humerus; 7, radius; 8, ulna; 9, radial carpal (radiale); 10, ulnar carpal (ulnare); 11, carpometacarpus; 12, first phalanges, digits III and IV; 13, second phalanx, digit II; 14, pelvic girdle; 15, ilium; 16, ischium; 17, pubis; 18, femur; 19, patella; 20, tibiotarsus; 21, fibula; 22, tarsometatarsus; 23, first metatarsal; 24, hypotarsus; 25, first phalanx, second digit; 26, first phalanx, third digit; 27, second phalanx, third digit; 28, third phalanx, third digit; 29, second phalanx, second digit; 30, third (distal) phalanx, second digit; 31, cervical vertebrae; 32, atlas; 33, axis; 34, sixth cervical vertebra; 35, thoracic vertebrae; 36, synsacrum; 37, free caudal vertebrae; 38, pygostyle; 39 and 40, vertebral ribs with uncinate processes; 41, sternal rib; 42, keel of sternum. (From Chamberlain, 1943. *Atlas of avian anatomy.* Courtesy Michigan Agricultural Experiment Station.)

the **calcaneum.** Medial to the calcaneum is the **astragulus (talus),** which articulates with the tibia and fibula. Distal to the astragulus is the **navicular,** and distal to it and the calcaneum is a row of four **cuneiform bones.** The sole of the foot is composed of five long **metatarsals,** and distal to them are the toes or **phalanges.**

THE AVIAN SKELETON

Birds are descendants of reptiles and share many skeletal features with their saurian predecessors. However, the evolution of the avian skeletal system has been molded by strong selection forces for development of the lightness and strength associated with flight; and because of the many morphological and physiological restrictions imposed by flight, birds constitute a vertebrate class morphologically more uniform within itself than perhaps any other group of the vertebrates. There are, however, major differences among the various avian groups, and these are generally associated with the invasion of new adaptive zones. For example, aquatic birds have developed specialized legs and feet for swimming and diving (e.g., loons and ducks), or wading (e.g., herons and ibises); whereas semiterrestrial birds have developed legs and feet suited to running (e.g., quail and turkeys).

The major features that distinguish the avian skeleton (**Fig. 5-13**) from that of most other vertebrates are those associated with flight, which lighten and strengthen the skeleton. This evolutionary trend has been accomplished by tremendous fusion and comcomitant disappearance of many bones. Fusion is most obvious in the areas of the sternum, pelvic girdle, and the fused thoracic (**notarium**) and sacral (**synsacrum**) areas of the vertebral column; disap-

pearance of bones is apparent especially in the wings and feet. Lightening of the skeleton is also accomplished by the extension of an **air-sac system** into many bones; this air-sac system replaces the marrow in many limb bones, but also may invade parts of the skull, the girdles, and the vertebral column.

Another feature unique to birds is the fusion of from four to eight embryonic caudal vertebrae to form a single bone, the **pygostyle,** to which the tail feathers are attached. The pygostyle also partitions a pair of **oil glands** (uropygial glands), which are important in waterproofing the feathers, particularly in swimming forms. *Archaeopteryx,* the first known bird, had a long series of free caudal vertebrae, with a pair of tail feathers attached to each. **Uncinate processes,** which lend rigidity to the rib cage, were not present in *Archaeopteryx.*

Flight · Flight is made possible by a unique structuring of the muscles and bones of the sternum and pelvic girdle. The sternum is an extensive bone exhibiting a ventrally directed **keel** or **carina** (L., *carina,* the keel of a ship) that serves as the bony surface for the attachment of the major flight muscles, the **pectoralis** and **supracoracoideus.** Certain of the flightless birds, including the ostrich, rhea, cassowary, emu, and kiwi, are collectively termed the **ratites** (L. *ratis,* raft), referring to the absence of a keel. The three bones of the pelvic girdle, the scapula, coracoid, and clavicle (two fused to form the **furculum**), meet to form a **triosseal canal** through which a tendon passes from the supracoracoideus muscle to insert on the posterior edge of the humerus. This muscle acts as an important part of the flight mechanism by elevating the humerus and wing. The pectoralis muscle inserts on the humerus and acts to pull the wing down.

6

The Musculature

The muscular system is perhaps the most difficult to study from the phylogenetic standpoint, of all the vertebrate structural systems. Modifications such as the splitting or fusion of muscle masses, the fusion of attachments, and the movement of appendages (e.g., in some teleosts, the movement of pelvics anterior to pectorals) make tracing muscle homologies throughout the various vertebrates extremely difficult, if not impossible in many cases. Muscle homologies and evolutionary trends are perhaps best established by embryonic origins and nervous innervations, but even this procedure is questionable at times. This chapter will attempt to trace a muscle system that is better understood than most from the phylogenetic standpoint, namely, the branchiomeric system; but first, a categorization of major vertebrate muscle systems is presented. **Table 6-4,** at the end of the chapter, shows the homologies between branchiomeric muscles of elasmobranchs and tetrapods.

Differentiation of the Major Masses of Mesoderm

After the mesodermal cells have become situated between the outer ectoderm and inner endoderm in the developing vertebrate embryo, three major regions of mesoderm may be distinguished (**Fig. 6-1**). There is a dorsal **epimere,** an intermediate mass known as the **mesomere** or **nephrotome,** and a ventral **hypomere** or **lateral plate.** The epimeres and mesomeres are metamerically arranged as a series of somites, while the hypomere is a solid sheet of tissue located laterally along the body. Each epimeric block becomes further differentiated into three regions. The lateral wall is known as the **dermatome;** it spreads beneath the epidermis to give rise to the dermis. The dorsomedial wall of the epimere is the **sclerotome;** it migrates beneath the notochord to give rise to most of the vertebral column. The ventromedial wall of the epimere is the **myotome;** it migrates both dorsally and ventrally to give rise to the striated skeletal muscle and the appendicular skeleton. Dorsally it gives rise principally to the epaxial muscles; ventrally, to the hypaxial mass. While the above events are occurring, the hypomere or lateral plate splits into two sheets. The outer layer or **somatic mesoderm** forms the outer lining of the body cavity, the **parietal peritoneum.** The inner layer or **splanchnic mesoderm** forms the **visceral peritoneum** and **dorsal** and **ventral mesenteries** that suspend the visceral organs within the body cavity. The splanchnic mesoderm and the adjacent endoderm of the **archenteron** or gut is referred to as the **splanchnopleure;** the somatic mesoderm and

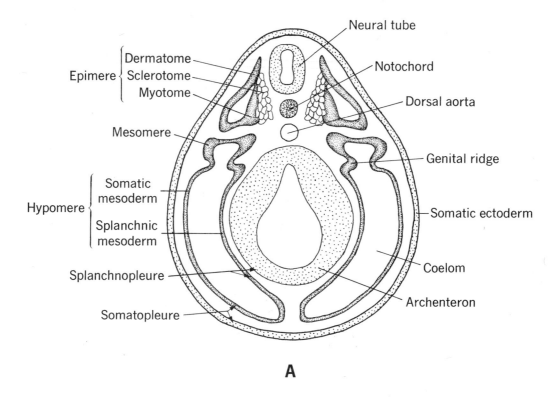

A

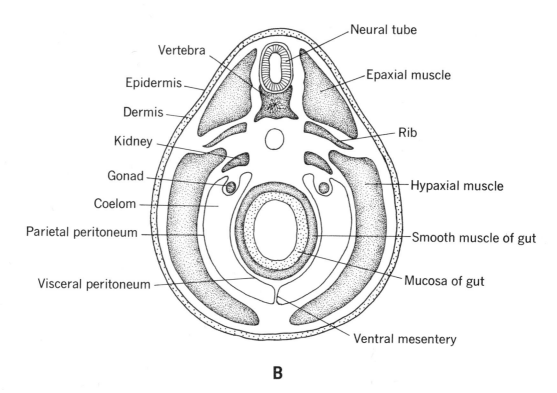

B

Fig. 6-1. Diagrammatic cross sections of a hypothetical vertebrate, illustrating the major embryonic masses of mesoderm (*A*), and some of their adult contributions (*B*).

Table 6-1. Embryonic contributions to the major mesodermal structures of the adult.

Embryonic Structures	Areas of Mesoderm	Adult Structures
Epimere	Dermatome	Dermis
	Sclerotome	Vertebrae
	Myotome	Epaxial and hypaxial muscle
		Appendicular skeleton and muscle
		Excretory organs
Mesomere		
		Reproductive ducts
Hypomere	Somatic "outer" layer	Parietal peritoneum
	Splanchnic "inner" layer	Visceral peritoneum
		Mesenteries
		Smooth muscle of viscera
		Gonads
		Heart, blood vessels, and blood cells
	Coelom	Coelomic body cavities

adjacent ectoderm, as **somatopleure.** Splanchnic mesoderm also gives rise to the smooth muscle of the gut, heart, and most of the gonads. The space between the two layers of the lateral plate is known as the coelom. The mesomere or nephrotome gives rise to the kidneys and the excretory ducts.

The contributions of the major masses of embryonic mesoderm structures are given in **Table 6-1.**

Major Muscle Masses

Major embryonic masses are illustrated in **Fig. 6-2.** As with many vertebrate systems, the muscular system may first be subdivided as somatic (parietal or myotomal) or visceral in character. **Somatic muscles** are those of the vertebrate embryo's "outer" tube, and are developed embryologically from myotomes. **Visceral**

muscles are derived primarily from the embryonic hypomere or lateral plate, and are of the "inner" tube (gut and pharynx); they are typically involuntary and of the smooth histological type, though the striated branchiomeric musculature is an exception. Somatic musculature constitutes the locomotor musculature, and is voluntary and striated. The heart is a further subdivision of the body musculature, but it is typically discussed as part of the circulatory system.

The somatic musculature may be further subdivided. **Axial muscles** (those along the axis of the body) comprise three major groups. The first group contains the **extrinsic eyeball muscles,** which develop from the three anterior head somites. The second group contains the **myotomal branchial muscles,** which develop from occipito-cervical myotomes; these are split (particularly in tetrapods) into a dorsal, **epibranchial** portion, part of the neck musculature, and a **hypobranchial** portion, which serves as the musculature of the throat and tongue. The

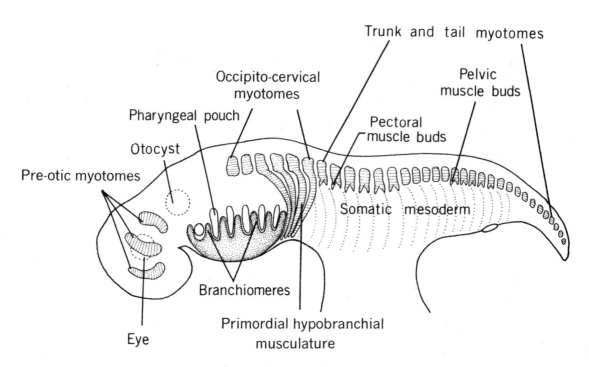

Fig. 6-2. Diagram showing the primordia of muscles in the vertebrate embryo. (After Torrey, *Morphogenesis of the vertebrates,* 3rd ed. Courtesy John Wiley & Sons, Inc., New York.)

hypobranchial muscles are innervated by the hypoglossal nerve (cranial nerve XII), which migrates ventrally and anteriorly beneath the pharynx, with the myotomes destined to become the hypobranchial muscles. The third group contains the **epaxial** and **hypaxial** musculature (derived from myotomes), which lie along the entire trunk, dorsal and ventral, respectively, to the **horizontal septum.** The epaxial and hypaxial muscles are innervated, respectively, by dorsal and ventral rami of the spinal nerves. These two muscle sets constitute the primary somatic musculature of vertebrates. **Appendicular muscles** are striated voluntary muscles developed embryologically from mesenchyme (initially from myotomes) lying within primordial limb buds. This musculature primitively consists of dorsal and ventral masses of muscles for operating fins; eventually the masses become subdivided into the numerous muscles of the tetrapod limb.

Aside from the smooth muscle of the gut and digestive system, the **branchiomeric muscles** are also visceral in origin. Though at least partially voluntary—and unlike other visceral muscle, striated in histology—this set of muscles is clearly visceral in origin, as evidenced by its innervation by the visceral motor neurons of cranial nerves. These muscles operate the visceral arches and their derivatives. Early in its evolution the branchiomeric musculature departed from the smooth histological type and became striated. This was probably associated with the vigorous primitive visceral functions of the pharynx: food-getting and respiration.

Structure of Striated Muscle

Striated muscle (**Fig. 6-3**) is voluntary muscle that forms the skeletal muscle of the body, and is derived primarily from the embryonic myotomes. The cells are elongate, cylindrical bodies that are multinucleated and have a char-

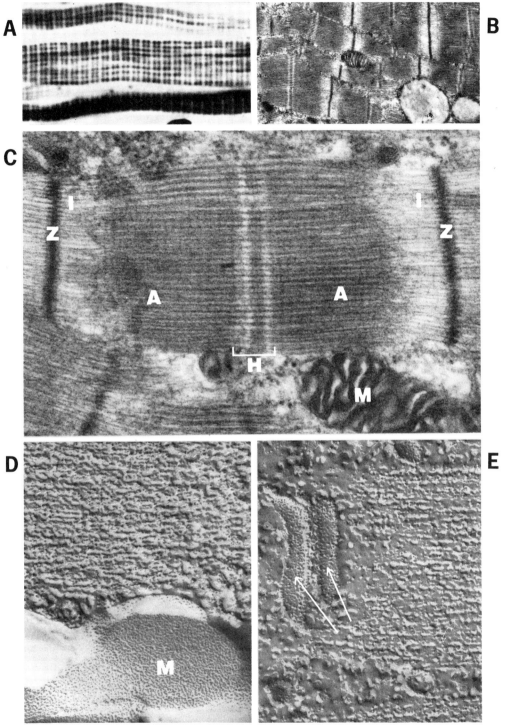

Fig. 6-3. Vertebrate striated muscle as seen by three different types of microscopy. *A,* light micrograph of muscle fiber composed of numerous myofibrils, 1800×; *B,* electron micrograph of sectioned muscle showing several myofibrils, 24,000×; *C,* electron micrograph of a single sarcomere and a mitochondrion (*M*) at lower right, 70,000×; *D,* electron micrograph of a freeze-etch preparation of a portion of one sarcomere and part of an adjacent mitochondrion, 70,000× (the freeze-etch technique involves a replication of the tissue with platinum and carbon, so not as much detailed structure can be seen as in *C*); *E,* electron micrograph of freeze-etch preparation of portion of one sarcomere and two tubular elements of T-tubule system (arrows), 70,000×. (Photos courtesy of Drs. Catherine Henley and D. P. Costello.)

acteristic striated appearance. In contrast, smooth muscle is composed of small, spindle-shaped cells with a single, centrally located nucleus, and is derived from embryonic mesenchyme. Smooth muscle is involuntary and contracts slowly, but does not fatigue quickly as striated muscle does. Cardiac muscle is a specialized striated muscle found only in the heart, but, like smooth muscle, derived from mesenchyme. It is a complex interwoven network of mononucleated cells that resembles striated muscle, but exhibits prominent cross bands, the intercalated discs, which separate the cells and provide a close functional and structural connection between them.

Striated muscle cells are covered on the external surface by a thin membrane, the sarcolemma, and are composed of muscle fibrils, functional units of which are termed sarcomeres. Individual sarcomeres may be recognized as the regions between Z-bands, the stable transverse septa. The contractile material of the sarcomere is composed of a series of partially overlapping filaments of actin (thin filaments attached to the Z-bands) and myosin (thick filaments), which forms the myofribrils and gives rise to the distinctive

microscopic band pattern. Upon contraction (according to the sliding filament model) the filament length remains constant, but the overlapping filaments slide past one another. As a consequence the actin is drawn into the myosin filament conformation (which is the strongly birefringent, anisotropic or A-band region). The light-colored, weakly birefringent I-bands (isotropic) are regions where no myosin is present, but only actin. The H- (hyaline) region of the A-band is the section where the actin does not extend into the A-region. The muscle contraction is set into action by the nerve action potential, which invades the myofibril via the membranes of the T-tubules (transverse tubules). These are in close contact with the terminal cisternae, reservoirs of the sarcoplasmic reticulum containing large amounts of calcium ions. The action potential, moving through the T-tubules to the cisternae, causes a change in the potential of the cisternal membranes, and a sharp increase in the permeability of the membrane to calcium. This sudden release of Ca++ into the intracellular medium is the direct stimulant for muscle contraction. Relaxation occurs when the calcium pump in the cisternal membranes returns the Ca++ to the cisternae.

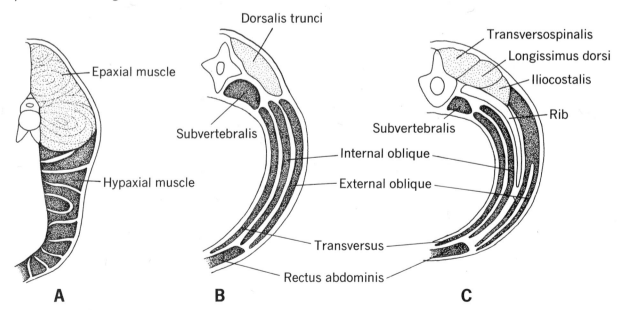

Fig. 6-4. Diagrammatic cross sections through the trunk regions of a shark (A), a salamander (B), and a lizard (C), showing the changes in the epaxial and hypaxial musculature.

Epaxial, Hypaxial, and Appendicular Musculature

Obtain a specimen of *Squalus* for study.

In the lamprey there is no distinction between epaxial and hypaxial portions of the musculature; however, in *Squalus* a clear distinction exists between musculature above the horizontal skeletogenous septum (epaxial) and musculature beneath the septum (hypaxial) (**Figs. 6-4,** and **6-5**). In elasmobranchs a por-

tion of the anteriormost hypaxial musculature has migrated beneath the pharynx to form the hypobranchial musculature. These muscles lie in an area between the coracoid bar and the mandible. Skin a portion of the trunk region, and observe that, as in the lamprey, the muscles occur as a series of metamerically arranged myomeres separated from each other by connective-tissue partitions known as myocommata or myosepta. However, unlike the case of the lamprey, the muscles of *Squalus* are composed of a dorsal epaxial bundle and a ventral

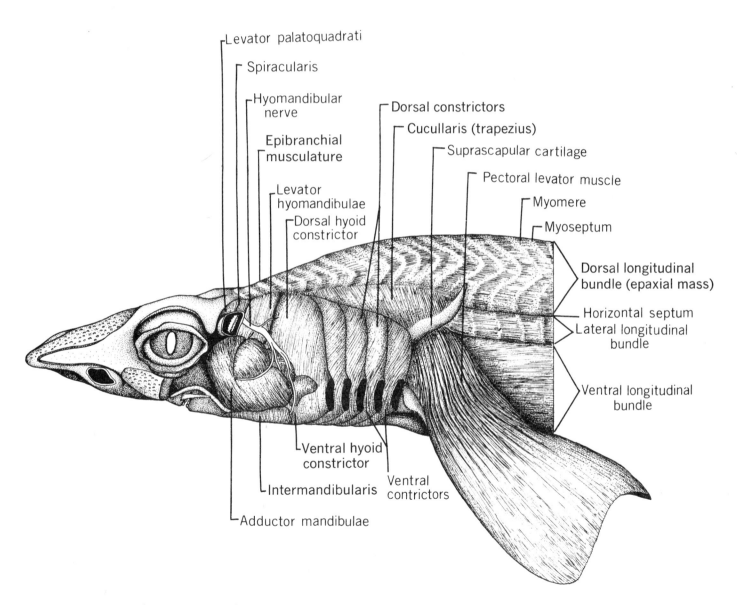

Fig. 6-5. Lateral view of the anterior musculature of *Squalus*. (Drawn from specimens and following Marinelli and Strenger.)

hypaxial bundle. Along the midventral line there is a white band of connective tissue, the **linea alba,** which persists throughout all of the vertebrates; it separates the myomeres of the two body sides. Appendicular musculature is a characteristic of fishes above the agnathostome level of organization. These muscles are developed from mesenchyme associated with the embryonic myotomes that invades the developing limb buds. Appendicular muscles are generally classified as **extrinsic** (extending from the body wall to the appendage), or **intrinsic** (extending from one part of an appendage to another part). In elasmobranchs the extrinsic muscles are fairly well defined, but the intrinsic muscles are rather poorly developed. Remove the skin from one of the pectoral fins (or examine a demonstration) and distinguish between the extrinsic appendicular abductors and adductors. Try to determine how the fin is operated.

Now turn to the mud puppy (**Fig. 6-6**), and note that the musculature still consists of myomeres separated by myocommata or myosepta, but the epaxial musculature is now structured to form a dorsal, longitudinal muscle bundle (called the **dorsalis trunci** or **longissimus dorsi**), and a deep series of intersegmental muscle bundles known as the **interspinalis** muscles; each passes from one vertebra to the next,

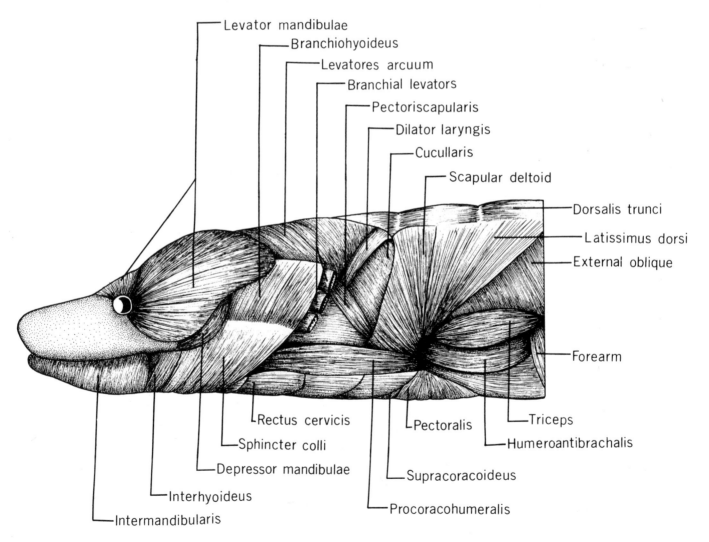

Fig. 6-6. Lateral view of the anterior musculature of *Necturus*.

originating on the posterior zygapophysis of one vertebra and inserting on the neural arch of the next posterior one. Aside from these specializations, the epaxial musculature is somewhat reduced in relative size compared to that of the fish.

Now turn to the hypaxial musculature, and determine that there is an outer **external oblique** muscle, an inner **internal oblique** muscle (whose fibers run opposite to those of the former), and an internal **transversus** (with more vertically directed fibers). Note also the **linea alba** on the midventral line, and the **rectus abdominis** (not present in *Squalus*) on either side of the linea alba. Note the great advancement in the limb musculature over that observed in *Squalus*. Dissect several extrinsic and several intrinsic appendicular muscles and try to ascertain their functions.

Skin a cat sufficiently to observe the musculature. From your experience with *Necturus* you should now be able to dissect the abdominal muscles of the cat; though there is some advancement over the musculature observed in *Necturus,* the muscles are similar and the nomenclature is the same. In advanced tetrapods the epaxial musculature is greatly reduced, and is generally confined to the immediate area of the vertebral column.

Notice the tremendous increase in complexity of the appendicular musculature over that observed in the amphibian. You should now skin and clean one side of a pectoral appendage so that you can dissect several extrinsic and several intrinsic appendicular muscles as indicated in **Table 6-2.** The **origin** of a muscle is the more stable end; the **insertion,** the opposite. In limb muscles the origin is proximal and the insertion distal. Identical muscles from opposite sides of the body may fuse in a middorsal or midventral ridge of connective tissue known as a **raphe.**

Table 6-2. Appendicular muscles (Figs. 6-7 and 6-8) and their origins, insertions, and actions.

Category	Muscle	Origin	Insertion	Action
EXTRINSIC				
A. ventral	1. **pectoantibrachialis** (most superficial pectoral muscle)	manubrium of sternum	proximal portion of ulna	adducts arm (pulls arm toward median plane)
	2. **pectoralis major** (caudal to the above)	cephalic region of sternum, and midventral thoracic raphe	along humerus (deltoid ridge to lower pectoral ridge)	draws arm inward (adducts forelimb)
B. dorsal	1. **acromiodeltoid**	acromion of scapula	proximal area of humeral shaft	flexes and rotates humerus
	2. **spinodeltoid**	scapular spine	proximal end of humerus	flexes and rotates humerus
INTRINSIC				
A. flexor	1. **brachialis**	lateral surface of humerus	ulna (distal to semilunar notch)	flexes antebrachium
B. extensor	1. **triceps brachii**	shaft and deltoid of humerus; glenoid border of scapula	olecranon of ulna	extends antebrachium

Branchiomeric Musculature

The muscles that operate the jaws and successive visceral arches are termed the branchiomeric muscles; they are visceral in origin, but are striated, and voluntary. Functionally, they serve to operate the jaws, and for breathing in fishes. They have become transformed to perform multifarious roles in the different vertebrates. In order to study the branchiomeric musculature in *Squalus*, *Necturus*, and the cat, the skin should be removed carefully on one anterior side of each animal as illustrated in **Figs. 6-5, 6-6,** and **6-7.** Beginning with the first visceral arch and proceeding posteriorly, dissect the muscles of each arch in each of the three organisms. Follow **Table 6-3** in making the dissections.

The first visceral arch in *Squalus*, *Necturus*, and the cat is concerned with the jaw mechanism. In **Squalus** there is a large **adductor mandibulae**, which inserts, along with the **intermandibularis**, on Meckel's cartilage; both raise the lower jaw. The **levator palatoquadrati** aids in raising the upper jaw. All first arch muscles are innervated by cranial nerve V (trigeminal). In *Necturus* also there is a large **levator mandibulae** that, along with the **intermandibularis**, helps to raise the lower jaw. In the cat the **adductor mandibulae** has become subdivided into separate **masseter** and **temporalis** muscles, which elevate the lower jaw. The in-

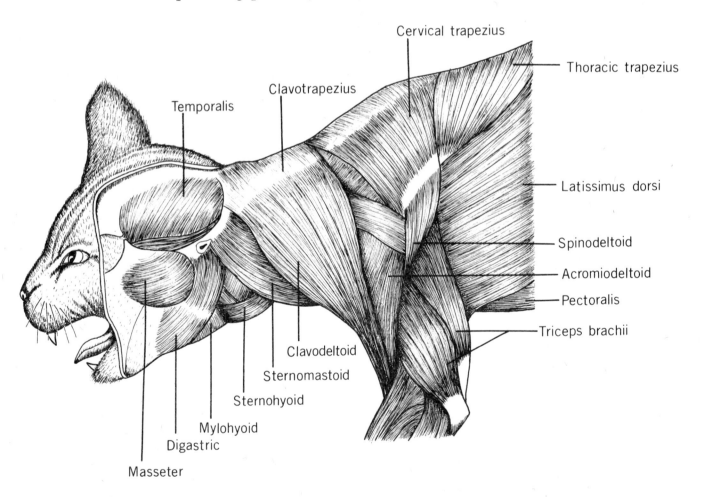

Fig. 6-7. Lateral view of the anterior musculature of the cat. (Drawn from specimens and following Walker.)

Table 6-3. The branchiomeric musculature of *Squalus*, *Necturus*, and *Felis*.

Branchiomeric muscle of arch I	Origin	Insertion	Action
Squalus			
adductor mandibulae	post. part of palatoquadrate cartilage	Meckel's cartilage	elevates lower jaw (closes jaw)
levator palatoquadrati [1]	otic capsule	palatoquadrate cartilage	elevates palatoquadrate cartilage
intermandibularis	midventral raphe of mandible	Meckel's cartilage	elevates floor of mouth, forcing water out of gill pouches
Necturus			
levator mandibulae	parietal and frontal bones	mandible	elevates lower jaw (closes mouth)
intermandibularis	midventral raphe of mandible	lower jaw	tenses floor of mouth
Felis [2]			
masseter	zygomatic arch	masseteric fossa and proximate parts of mandible	elevates lower jaw
temporalis	temporal fossa	inner and outer surfaces of coronoid process of mandible	elevates lower jaw
mylohyoid [3]	mandible	median raphe and hyoid	raises floor of mouth and pulls hyoid forward
digastric [4]	occipital bone lateral to condyle and mastoid region of temporal complex	ventral surface of mandible	depresses lower jaw (opens mouth)
tensor tympani [1]	medial wall of tympanic cavity	malleus	adjusts malleus to sound-wave intensity
pterygoideus [1]	base of skull	medially on mandibular ramus	aids in closing jaws

[1] Impractical to dissect.

[2] Muscles of *Felis* are extremely similar to those of other mammals, including man.

[3] Note that the muscles lying deep to the mylohyoid and surrounding area constitute a prehyoid portion of hypobranchial myotomal musculature.

[4] Partially derived (posterior portion) primitively from the interhyoideus of arch II; hence innervated by cranial nerves V and VII.

Table 6-3. The branchiomeric musculature of *Squalus, Necturus,* and *Felis.* (cont.)

Branchiomeric muscle of arch II	Origin	Insertion	Action
Squalus			
levator hyomandibulae	otic capsule	hyomandibular cartilage	compresses gill pouch
interhyoideus (deep to intermandibularis)	midventral raphe	ceratohyal cartilage	compresses gill pouch
dorsal and ventral hyoid constrictors		(continuation of interhyoideus)	

(Note the other branchiomeric gill constrictors of arches III–VI.)

Necturus			
epihyoideus		(forms stapedius muscle, but may be secondarily reduced or absent in salamanders where middle ear is reduced, as in *Necturus*)	
interhyoideus (sphincter colli or gularis, particularly posterior portion)	hyoid arch medial to jaw angle	midventral raphe	tenses throat
depressor mandibulae	1st branchial arch	posterior part of mandible	opens jaw
branchiohyoideus (larger, posterior portion of d. mandibulae)	1st branchial arch	ceratohyal cartilage	draws hyoid backward
Felis			
stapedius (not to be dissected)	medial wall of tympanic cavity	stapes	adjusts stapes to sound-wave intensity
stylohoid (lies lateral to post. portion of digastric)	stylohyal process	anterior horn of hyoid	aids in swallowing
digastric		(see under arch I derivatures—posterior portion is from arch II)	
platysma and mimetics [5]	fascia of neck and head region	skin of neck and head region	moves skin for facial expression

[5] These muscles are derived from the sphincter colli; functionally similar integumentary muscles of the trunk (cutaneus maximus or panniculus carnosus) are myotomal in origin.

termandibularis is now termed the **mylohyoid.** In addition, part of the intermandibularis gives rise to the anterior region of the **digastric** which serves to depress the lower jaw. There is a **tensor tympani,** which puts tension on the ear drum, and a **pterygoideus,** which aids in closing the jaws.

Branchiomeric muscle of the second visceral arch, the hyoid arch, is innervated by cranial nerve VII, the facial. In *Squalus,* the **interhyoideus** (deep to the intermandibularis) compresses the gill pouch, and the **levator hyomandibulae** (from the otic capsule to the hyomandibular cartilage) also acts as a con-

Table 6-3. The branchiomeric musculature of *Squalus, Necturus,* and *Felis.* (*cont.*)

Branchiomeric muscle of arches III–VI

In *Squalus* the visceral cartilages of these arches contain branchiomeric muscle mainly in the form of constrictors (which you should locate in your specimens). In the evolution of the higher fishes and tetrapods, with a reduction in the number of gills and the posterior gill region, much of the posterior branchiomeric musculature is lost. However, some becomes associated with the larynx as the intrinsic laryngeal musculature and the wall of the pharyngeal region. The major tetrapod muscles, which include in mammals the stylopharyngeus, thyroarytenoideus, cricoarytenoideus, and cricothyroideus, will not be dissected here, but may be observed collectively on the ventral laryngeal surface of the cat.

Branchiomeric muscle of arch VII	Origin	Insertion	Action
Squalus			
cucullaris (the united levators of the posterior five gill arches, lying dorsal to the superficial constrictors)	dorsal fascia and surface of epibranchial musculature	epibranchial cartilage of last visceral arch and ventral scapula	elevates gill arches and scapular cartilage
Necturus			
cucullaris (see note above)	dorsal cervical fascia	scapula (near the glenoid cavity)	draws the scapula anteriorly and dorsally
Felis			
branchiomeric muscle derived (at least partially) from cucullaris:			
clavotrapezius	lambdoidal ridge, and midventral raphe over axis vertebra	clavicle and raphe between clavotrapezius and clavobrachialis	pulls scapula outward and forward
sternomastoid	midventral raphe of thorax and sternal manubrium	mastoid region of temporal complex, and lambdoidal ridge	turns head; pulls head toward neck
cleidomastoid	mastoid region of temporal complex	clavicle and associated raphe	essentially same as sternomastoid

strictor. Dorsal and ventral **hyoid constrictors** extend anteriorly to insert on the mandible. In *Necturus* the **interhyoideus** (called the **sphincter colli** posteriorly) is present; in addition, the large **depressor mandibulae** is the primary muscle involved in opening the jaws. It inserts on the posterior end of the mandible. The **branchi-** ohyoideus goes forward beneath the interhyoideus and intermandibularis to insert on the ceratohyal cartilage; it serves to draw the hyoid arch backward. In the cat there is a **stapedius** muscle, which is within the middle ear cavity and acts on the stapes; it is impractical to dissect. **Platysma** and **mimetic** muscles are present superficially and serve for facial expression. The major hyoid arch muscles, however, are the **stylohyoid** and posterior half of the **digastric.** The digastric depresses the lower jaw. The stylohyoid lies lateral to the posterior belly of the digastric. It inserts on the body of the hyoid and aids in swallowing.

The branchiomeric muscle of the third visceral arch is innervated by cranial nerve IX (glossopharyngeal), that of the succeeding arches by cranial nerve X (vagus). In *Squalus*

the branchiomeric muscle of arches III and IV is in the form of dorsal and ventral constrictors, which function to expand and compress the gill pouches. The **cucullaris** (possibly derived from many of the posterior arches, including VII) elevates the gill arches and the scapular cartilage. In *Necturus* there are branchial levators on the external gills, and a series of **levatores arcuum** dorsal to the bases of the gills. The **cucullaris** in *Necturus* draws the scapula anteriorly and dorsally. In the cat much

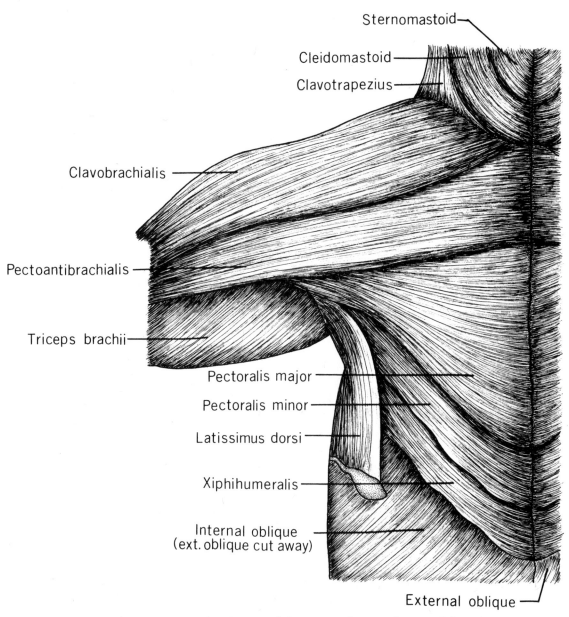

Fig. 6-8. Superficial view of the pectoral musculature of the cat.

of the posterior branchiomeric musculature has been lost, but some remains as the intrinsic muscles of the larynx. In addition there are several muscles associated with the shoulder region. These include the **clavotrapezius**, the **sternomastoid**, and the **cleidomastoid** (Figs. 6-7 and 6-8). For comparison, selected branchiomeric muscles of a monkey are shown in **Fig. 6-9.**

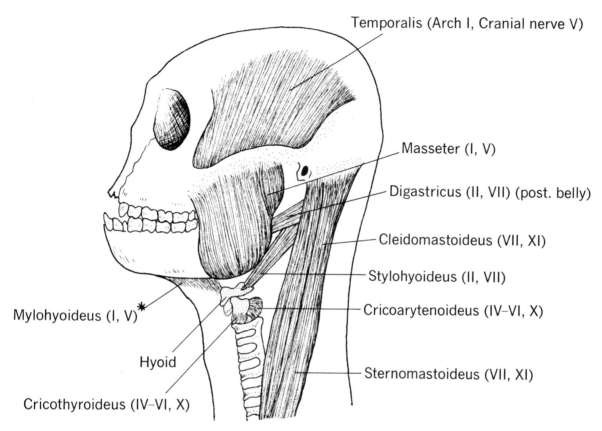

Fig. 6-9. Selected branchiomeric muscles of a monkey. After each muscle name is given the visceral arch from which it is derived and the cranial nerve innervation. * The posterior part of the mylohyoideus is derived from visceral arch II and is predictably innervated by cranial nerve VII.

Table 6-4.   Homologies between branchiomeric muscles of elasmobranchs and tetrapods.

	Primitive skeletal elements	Muscles	
		Elasmobranchs	Tetrapods
Mandibular arch [1] (visceral arch I) Innervation: cranial nerve V (trigeminal)	palatoquadrate (dorsal cartilage, or its ensheathing bones)	levator maxillae superioris	
	Meckel's cartilage (ventral cartilage or its ensheathing bones)	adductor mandibulae	adductor mandibulae or derivatives thereof: (a) masseter (b) temporalis
	Meckel's cartilage	intermandibularis	intermandibularis or derivatives thereof: (a) mylohyoideus (b) ant. belly of digastricus
Hyoid arch [1] (visceral arch II) Innervation: cranial nerve VII (facial)	hyomandibular	levator hyomandibulae	stapedius
	ceratohyal	constrictors	stylohyoideus
	basihyal	interhyoideus	interhyoideus post. belly disgastricus depressor mandibulae sphincter colli (= gularis) or derivatives thereof: (a) platysma (b) mimetics
Visceral arch III (first true gill arch in elasmobranchs) Innervation: cranial nerve IX (glassopharyngeal)	gill arches	constrictors	stylopharyngeus
Visceral arches IV–VI (gill arches) Innervation: cranial nerve X (vagus)	gill cartilages	constrictors	laryngeal musculature
Visceral arch VII (gill or visceral cartilage) Innervation: occipitospinal nerves in anamniotes; XI in amniotes	gill cartilage, or, in *Squalus*, 7th visceral cartilages	cucullaris (also partly derived from constrictors of arches IV–VI)	trapezius sternomastoideus cleidomastoideus

[1] Primitively, gill cartilages with branchiomeric constrictor and adductor muscles.

7

The Nervous System and Sense Organs

The nervous system is divided conveniently into a **central nervous system (CNS)** and a **peripheral nervous system.** The CNS includes the brain and spinal cord, both of which are hollow. Embryologically, there is a dorsomedial thickening of ectoderm along the longitudinal axis of the embryo forming the neural plate of neural ectoderm (**Fig. 7-1**); then a medial depression occurs, and a subsequent elevation of the margins, forming neural folds. When the folds unite dorsally the neural tube is formed. The neural crest cells differentiate along the margins of the folds. The various regions of the CNS are termed white or gray matter depending on their composition. Gray matter is composed of unmyelinated fibers and cell bodies of the neurons; most gray matter is located centrally, within the CNS. White matter is composed of myelinated fibers, and typically is located peripherally, at least in the spinal cord. In the telencephalon of the brain of higher vertebrates the composition of gray and white matter is reversed. However, in all forms the gray matter in the brain tends to be in the form of small patches of nuclei.

The peripheral nervous system is composed of the cranial and spinal nerves and the autonomic nervous system; it consists of the neurons (sensory and motor) that supply the receptors and effectors of the organism. The spinal nerves (**Figs. 7-2** and **7-3**) are metamerically arranged, each one exiting between successive vertebrae. Upon exiting, there is a joining of the dorsal and ventral roots of these nerves. Along the dorsal roots are located the ganglia containing the cell bodies of sensory neurons. The ventral, motor neurons have their cell bodies located within the CNS. After the dorsal and ventral roots join, they divide up into a **dorsal ramus** that goes into the epaxial region, a **ventral ramus** that goes into the hypaxial region, and one or more interconnecting communicating rami with neurons that send branches to a **visceral ramus.** The autonomic nervous system (**Fig. 7-4**) is the motor portion of the peripheral nervous system, which supplies the smooth muscle, cardiac muscle, visceral organs and glands; these neurons exit from the CNS via various spinal and cranial nerves. Cranial nerves are for the most part derived from spinal nerves, but many have lost much of their resemblance to the latter.

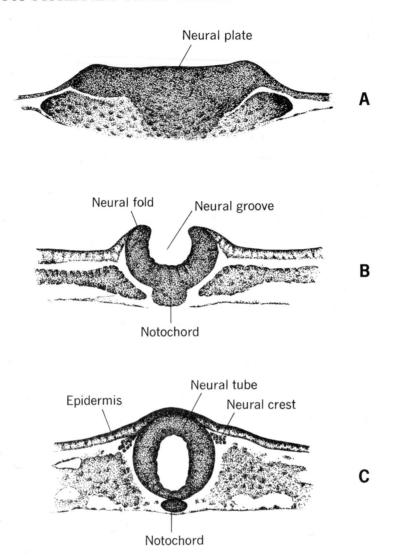

Fig. 7-1. Stages in the formation of the neural tube of the chick. *A*, a thickening and stratification in the dorsomedial neural ectoderm forms a neural plate; *B*, the middle of the neural plate buckles downward with a concomitant upward movement of the neural folds to form the neural groove; and *C*, the neural folds meet dorsally to form the neural tube, and a wedge of cells, the neural crest, appears between the neural tube and the epidermis. The neural crest cells form or induce the formation of such diverse structures as sensory neurons of cranial and spinal nerves, the adrenal medulla, certain nerve sheaths, melanophores, and parts of the branchial skeleton.

Fishes: Brain and Cranial Nerves

The basic architectural plan of the vertebrate brain (see **Table 7-1**) and cranial nerves is beautifully exemplified by *Squalus* (**Fig. 7-5**), and dissection is facilitated by the cartilaginous composition of the neurocranium. Obtain a specimen of *Squalus* and remove the skin, associated tissue, and cartilage from the dorsal aspect of the head and from around one eye until the entire dorsal area of the brain is exposed.

Locate the two **cerebral hemispheres** of the **telecephalon;** they send out the large **olfactory tracts,** which terminate in the olfactory bulbs. Cut into one of the olfactory sacs and observe the **lamellae** of the olfactory epithelium. The

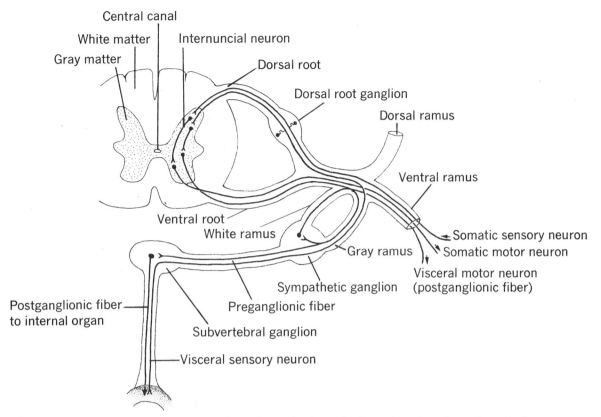

Fig. 7-2. Diagrammatic cross section through the spinal cord and a thoracic spinal nerve to illustrate the functional components of the nervous system. The major nerve components are as follows: **Somatic sensory (afferent)** fibers serve a proprioceptive function from receptors in striated muscle and tendons, and for sensations of temperature, pain, pressure, and touch from general cutaneous receptors. **Visceral sensory (afferent)** fibers convey messages from the viscera. **Somatic motor (efferent)** fibers go to the myotomal muscle of the body. **Visceral motor (efferent)** fibers constitute the autonomic nervous system, and send fibers to smooth muscle, cardiac muscle, and glands (see Fig. 7-4). The dorsal ramus contains fibers of both somatic sensory and somatic motor neurons.

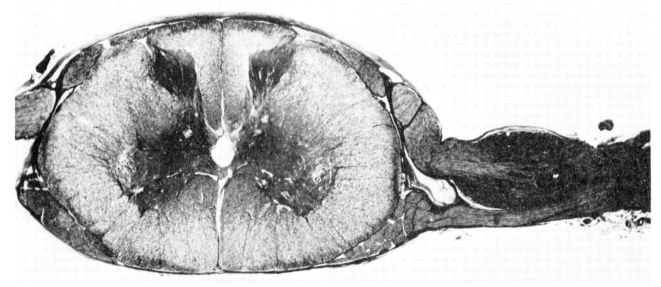

Fig. 7-3. Cross section of the spinal cord and a spinal nerve of a rabbit (see Fig. 7-2 for details of structure). (Photo courtesy of Drs. Catherine Henley and D. P. Costello.)

Table 7-1. The three major subdivisions of the vertebrate brain and their component parts.

PROSENCEPHALON (FOREBRAIN)

Telencephalon	*Diencephalon*
Lateral ventricles	Third ventricle
Rhinencephalon	Epithalamus
Olfactory bulbs, tracts, and lobes	Pineal body (epiphysis)
Cerebral hemispheres	
Corpora striata	Thalamus
Corpus callosum	Hypothalamus
Neocortex	Optic chiasma
Paraphysis	Infundibulum and pituitary gland
	Saccus vasculosus

MESENCEPHALON (MIDBRAIN)

Tectum (roof)
  Optic lobes
  Auditory lobes
Cerebral peduncles
Cerebral aqueduct

RHOMBENCEPHALON (HINDBRAIN)

Metencephalon	*Myelencephalon*
Cerebellum	Fourth ventricle
Pons	Medulla oblongata

Fig. 7-4. The human autonomic nervous system. Neurons of the autonomic nervous system (ANS) exit via four of the twelve cranial nerves, and approximately half of the 31 spinal nerves. The ANS is divisible into sympathetic and parasympathetic nervous systems; neurons of the former exit via spinal nerves in the thoracic and lumbar regions, of the latter from cranial nerves and spinal nerves in the sacral region. Neurons from both systems supply identical organs but act antagonistically. Both systems have pathways involving two motor (efferent) neurons; a presynaptic neuron (with cell bodies in the CNS) exits the CNS and synapses with a second, postsynaptic neuron, which innervates a particular target organ. In this two-fiber, peripheral relay system, the presynaptic neurons of the sympathetic system synapse with the postsynaptic neurons in a series of ganglia near the spinal cord (see Fig. 7-1), or in large ganglia in the abdominal cavity. Thus the second neuron is quite long, and because of a fiber ratio of 1 presynaptic to 15 postsynaptic neurons, the sympathetic system has a widespread effect. The transmitter substance for the presynaptic neurons at the sympathetic ganglia is Ach (acetylcholine); the postsynaptic transmitter substance at the target organs is either noradrenaline (norepinephrine) or adrenaline (epinephrine). The parasympathetic nervous system is characterised by long presynaptic neurons that synapse with the postsynaptic neurons in ganglia near or within the target organs; thus the postsynaptic fibers are short. This fact, combined with a pre- to postsynaptic fiber ratio of 1:2; gives the parasympathetic system a relatively less widespread effect. The transmitter substance for both pre- and postsynaptic neurons of the parasympathetic system is Ach. In general, the sympathetic nervous system produces an effect similar to that of the hormone of the adrenal medulla, in preparing the organism for "fight-or-flight" responses: the parasympathetic system has an opposing effect. (From Keeton, Fig. 10.11.)

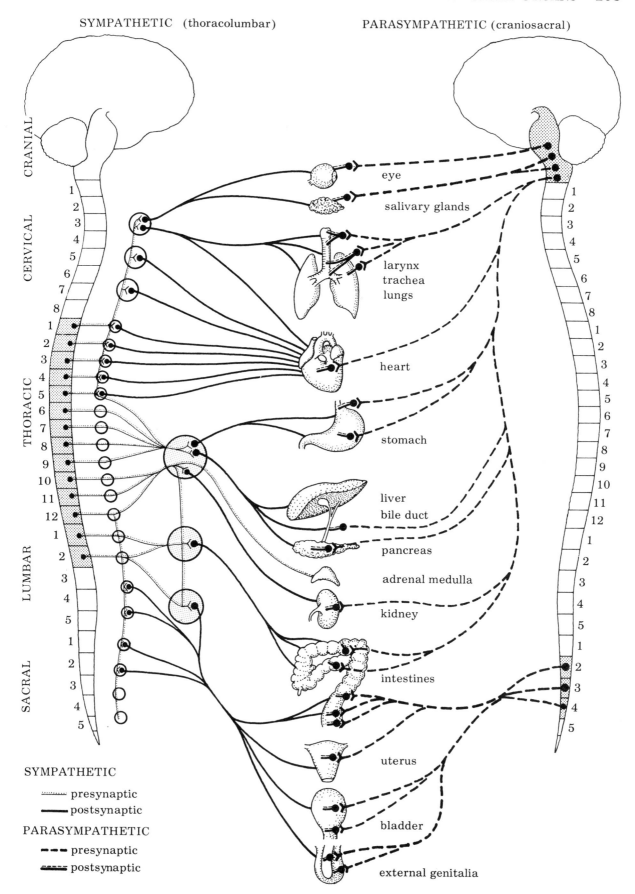

SYMPATHETIC (thoracolumbar)

PARASYMPATHETIC (craniosacral)

CRANIAL

CERVICAL

THORACIC

LUMBAR

SACRAL

eye

salivary glands

larynx
trachea
lungs

heart

stomach

liver
bile duct

pancreas

adrenal medulla

kidney

intestines

uterus

bladder

external genitalia

SYMPATHETIC

········· presynaptic

———— postsynaptic

PARASYMPATHETIC

– – – presynaptic

══ postsynaptic

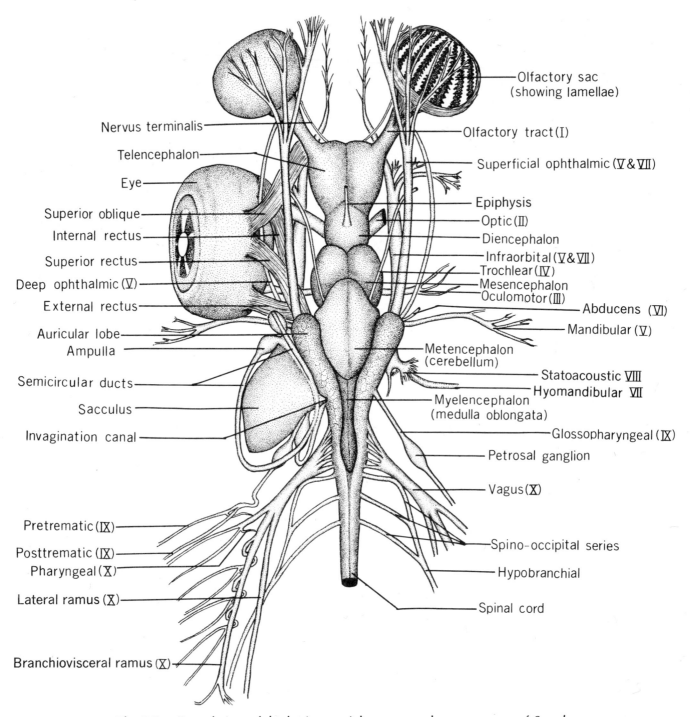

Fig. 7-5. Dorsal view of the brain, cranial nerves, and sense organs of *Squalus*.

cell bodies of the **olfactory nerve** (cranial nerve I) are located within this epithelium. The olfactory nerve carries somatic sensory impulses for olfaction from the olfactory sac to the brain; it is a purely sensory nerve. The olfactory lobes are at the point where the tracts enter the brain, but are difficult to distinguish from the hemispheres. The cavities within the two cerebral hemispheres are termed the lateral ventricles.

The smaller depressed area posterior to the hemispheres is the **diencephalon.** The **epiphysis** (vestigal **pineal body**) may be seen in the dorsomedial aspect of the **epithalamus,** a term for the roof of the diencephalon. The floor of the diencephalon is termed the **hypothalamus,** and the lateral walls, the **thalamus.** The very thin roof of the brain at the level of the diencephalon is the **tela choroidea;** it contains a choroid plexus, which secretes cerebrospinal fluid into the ventricles. The cavity of the diencephalon is termed the third ventricle.

The large **optic lobes** form the lateral walls of the **mesencephalon.** Posteriorly is the **metencephalon;** its dorsal aspect is the **cerebellum,** and its lateral flaps are termed the **auricles** or auricular lobes.

Posterior to the metencephalon is the elongate **myelencephalon.** This forms the greater part of the **medulla oblongata,** which is continuous with the spinal cord posteriorly. The roof of the medulla is a thin tela choroidea that forms the **posterior choroid plexus.** The cavity of the medulla is the fourth ventricle of the brain.

The ventral aspect of the brain will be studied after the cranial nerves.

CRANIAL NERVES

Expose each of the cranial nerves of **Squalus** (Fig. 7-5), and follow them to their target structures.

O. **Nervus terminalis.** This tiny nerve is located on the medial side of the olfactory tract, between the olfactory lobes and the sac. It probably serves for general sensory functions from the olfactory epithelium, but its precise function remains obscure.

I. **Olfactory.** The olfactory is a purely sensory nerve; it was discussed along with the telencephalon.

II. **Optic.** The optic, another purely sensory nerve, exits the brain from the ventral aspect of the diencephalon; it begins as the **optic chiasma,** and enters the eye. Its cell bodies are located within the retina. These cell bodies initiate sight impulses to the brain. Embryologically the optic nerve develops as an outgrowth of the brain, and in that sense is a brain tract rather than a true cranial nerve.

III. **Oculomotor.** The oculomotor arises from the ventral surface of the mesencephalon, but is difficult to observe. It may be observed as it supplies the internal and superior rectus eye muscles. It also supplies two more ventral extrinsic eye muscles. It serves as a motor nerve to four extrinsic eye muscles, and performs a sensory function, for proprioception from the muscles. In addition (though not observable) it has autonomic fibers leading to the constrictor muscles of the iris for partial control of the diameter of the pupil.

IV. **Trochlear.** This nerve originates from the dorsal aspect of the brain where the optic lobes connect with the cerebellum. The nerve passes over the optic lobes to supply the superior rectus extrinsic eye muscle. Its fiber composition is like that of III.

V is discussed with VII.

VI. **Abducens.** The abducens arises from the ventral aspect of the medulla and is best observed when the brain is removed. It supplies the external rectus extrinsic eye muscle with fibers similar to those of III and IV.

V. **Trigeminal.** Recall that this is the nerve of the mandibular arch. It also sends sensory fibers to the head region for reception of general cutaneous sensations. The trigeminal exits in a common root with the facial (VII) from the dorsolateral area of the medulla, slightly posterior to the auricles of the cerebellum. There are four major branches of this nerve; each will be covered separately.

1. **Superficial ophthalmic.** The large superficial ophthalmic passes with fibers from VII along the dorsal aspect of the orbit, and into the rostral region. It is a purely sensory branch, and supplies general cutaneous receptors in the head region.

2. **Deep ophthalmic (profundus).** The deep ophthalmic is a purely sensory branch of the trigeminal only; it passes through the orbit and is embedded in connective tissue on the

posterior part of the eyeball. It joins the superficial ophthalmic after exiting the orbit and serves a similar function.

3. **Maxillary.** This branch, along with the buccal ramus of the facial, forms the large infraorbital trunk, which is a very thick nerve extending anteriorly across the floor of the orbit. Its fibers pass to the upper jaw and rostrum where they serve a sensory function for general cutaneous receptors.

4. **Mandibular.** This branch is located in the posterior wall of the orbit. It may be traced to the mouth angle where it sends out both motor and sensory fibers, all from nerve V. It sends motor fibers to the muscles of the jaw (mandibular arch), which insert on Meckel's cartilage. It has sensory fibers for proprioception from the muscles, and for cutaneous receptors in the skin overlying the area of the mandible.

VII. **Facial.** The facial is the cranial nerve of visceral arch II, the hyoid arch. Superficial ophthalmic and buccal rami are associated with V and are sensory for the lateral line organs, and the ampullae of Lorenzini (special cranial sense organs). In addition, there is a large **hyomandibular branch.** It exits the brain in common with the trigeminal, and passes under the skin immediately posterior to the spiracle (see Fig. 6-5). This nerve sends visceral motor fibers to the hyoid muscles and the cephalic canal system and ampullae of Lorenzini, and has sensory fibers on the skin of arch II and also in the roof of the pharynx.

Before proceeding with nerve VIII, the point of common emergence of V and VII should be dissected out at the posteromedial corner of the orbit; it is there that the sensory cell bodies of V are located in a **trigeminal ganglion,** observable as a slight enlargement of the nerve.

VIII. **Auditory.** This is the third of the purely sensory cranial nerves; it contains only somatic sensory fibers from the various parts of the inner ear. This nerve arises immedi-ately posterior to V and VII on the lateral area of the medulla, and passes to the semicircular canals and associated structures of the inner ear. It is often termed the **stato-acoustic** in fishes.

IX. **Glossopharyngeal.** This nerve, which exits the lateral wall of the medulla immediately posterior to the eighth, is the nerve of the third visceral arch, the first true gill arch in *Squalus.* Before it divides into its three branches, there is a swelling, the **petrosal ganglion,** which contains its cell bodies. In order to trace the glossopharyngeal nerve it is necessary to cut into the tissue dorsal and ventral to the first external gill slit. By careful dissection one can locate and trace a **posttrematic** (mixed) branch that sends fibers down into the posterior demibranch of the first gill pouch. It sends motor fibers to the branchiomeric muscle and returns visceral sensory fibers from the posterior demibranch. A purely visceral sensory branch, the **pretrematic,** sends fibers down the anterior demibranch, and a **pharyngeal** (sensory) arises at the bifurcation of the post- and pretrematic, and goes into the roof of the pharynx.

X. **Vagus.** The vagus nerve (L., *vagus,* wandering) is the cranial nerve of the visceral arches posterior to the third. This large nerve arises from the posterolateral side of the medulla and proceeds caudally to give rise to several major branches. The **lateral ramus** is the longest nerve of *Squalus;* embedded within the muscle, it passes posteriorly to the tail, supplying the lateral line canal system along its entire length. The **branchiovisceral ramus** gives off branches to each of the last four gill slits; each have pretrematic, posttrematic, and pharyngeal rami that function similarly to those of IX. It then continues to the stomach and sends a branch to the pericardial cavity. This branch contains fibers of the autonomic nervous system, in addition to visceral sensory fibers. The branchiovisceral ramus also sends a small branch to the cucullaris muscle, in keeping with the pattern of innervation of the branchiomeric musculature.

XII. **Occipital nerves.** Immediately posterior to the vagus is a series of nerves termed the occipital nerves; they resemble spinal nerves that have lost their dorsal roots, which some authors believe are incorporated into the vagus. In amniotes these nerves will contribute to the formation of a cranial nerve, the hypoglossal. Posteriorly these nerves merge to form the **hypobranchial nerve,** which sends somatic motor fibers to the hypobranchial musculature, and also contains some sensory fibers.

VENTRAL VIEW OF THE BRAIN

Remove the brain by first cutting the medulla and the roots of cranial nerves. Study the features illustrated in **Fig. 7-6.**

Note the **optic chiasma;** its structure is such that it allows fibers from both optic nerves to go to both sides of the brain, permitting stereoscopic vision. Recall that the floor of the dien-

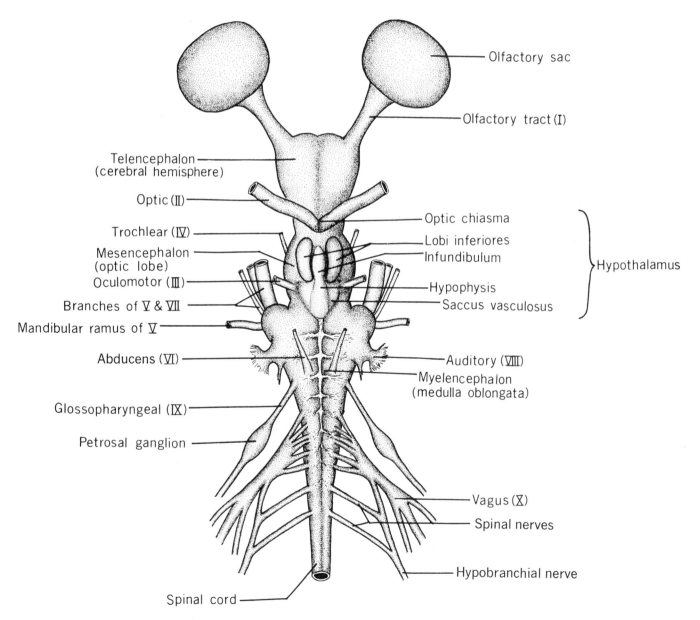

Fig. 7-6. Ventral view of the brain of *Squalus.*

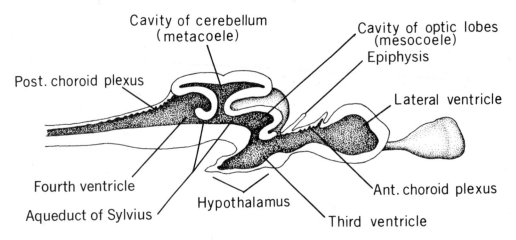

Fig. 7-7.   Sagittal section of the brain of *Squalus*, showing the ventricles.

cephalon is termed the hypothalamus. Examine the **infundibulum** with its lateral **lobi inferiores.** The **hypophysis (pituitary gland)** is located between the lobi inferiores and extends posteriorly. The **saccus vasculosus** lies posterior to the lobi inferiores. Examine the recess in the neurocranium that houses the hypophysis; it is termed the **sella turcica.**

By making a sagittal section of the brain, the major ventricles may be observed (**Fig. 7-7**). The **lateral ventricles** are within the two cerebral hemispheres; the **third ventricle,** within the diencephalon. The **fourth ventricle** is within the medulla. The third ventricle connects with the fourth via the **aqueduct of Sylvius.**

Mammals

It is far beyond the scope of this laboratory course to dissect the mammalian brain in all its detail. As was pointed out previously, the brain of *Squalus* exhibits the basic architectural plan of the vertebrate brain. The major difference observed in the mammalian brain is the tremendous overgrowth of the telencephalon to form the large cerebral hemispheres, which are thrust caudad over the posterior region of the brain. Thus if one removes the cerebral hemispheres and the hemispheres of the cerebellum, one encounters a brain typical of other vertebrates. With a few exceptions (to be pointed out

later), the cranial nerves are similar to those of *Squalus* except that there are twelve (within the foramen magnum) in amniotes. The mammalian cranial nerves have been studied previously in connection with the foramina of the cat skull; they will only be treated here as they are observable emerging from the brain.

The mammalian brain (**Figs. 7-8, 7-9, and 7-10**) and the emergent stumps of the cranial nerves may be studied using preserved sheep or cat brains, or the cat head may be sectioned sagittally and the halves of the brain removed by careful dissection. If you remove the brain you will have to leave the tough **dura mater,** which is the membrane covering the brain. There are other membranes (**meninges**) of the brain, but they are difficult to observe: the **pia mater** is a vascular tissue close to the surface of the brain, the **arachnoid membrane** is interposed between the dura and pia layers. Between the arachnoid membrane and the pia mater is the **subarachnoid space,** which is filled with cerebrospinal fluid to help cushion the brain.

TELENCEPHALON

The most salient feature of the mammalian brain is the presence of the tremendous **cerebral hemispheres** (known together as the **cerebrum**). On the surface are numerous folds, the **gyri,** which are bounded by grooves, the **sulci.**

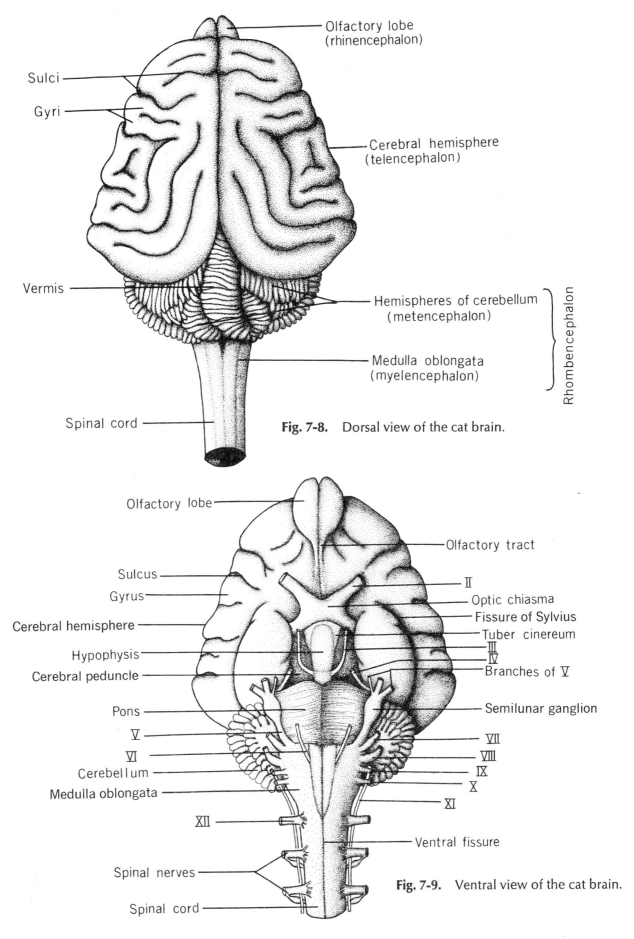

Fig. 7-8. Dorsal view of the cat brain.

Olfactory lobe (rhinencephalon)

Sulci

Gyri

Cerebral hemisphere (telencephalon)

Vermis

Hemispheres of cerebellum (metencephalon)

Medulla oblongata (myelencephalon)

Rhombencephalon

Spinal cord

Olfactory lobe

Olfactory tract

Sulcus

Gyrus

II

Optic chiasma

Fissure of Sylvius

Cerebral hemisphere

Tuber cinereum

Hypophysis

III

IV

Cerebral peduncle

Branches of V

Pons

Semilunar ganglion

V

VII

VI

VIII

Cerebellum

IX

Medulla oblongata

X

XI

XII

Ventral fissure

Spinal nerves

Fig. 7-9. Ventral view of the cat brain.

Spinal cord

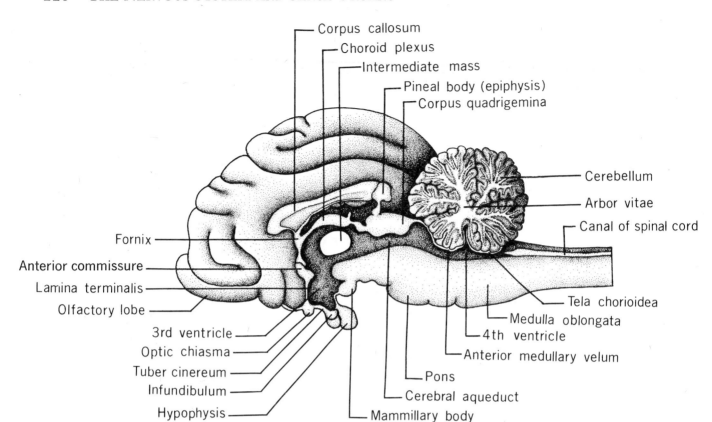

Fig. 7-10. Sagittal section of the cat brain.

In sagittal section or by spreading the dorsal hemispheres apart, one may observe the **corpus callosum**, the broad, white band of fibers that connects the two hemispheres. The corpus callosum extends ventrally as a similar band, the **fornix**. A lateral fissure or **fissure of Sylvius** divides the hemispheres into **frontal** and **temporal lobes.**

The **olfactory lobes** and olfactory tracts are present ventrally. The olfactory lobes are sometimes treated as a separate region of the brain, the **rhinencephalon.** They lie adjacent to the cribriform plate of the mesethmoid and receive the neurons from the olfactory epithelium, which in mammals is housed on the ethmoturbinals.

In primitive vertebrates the telencephalon was predominantly a center for olfaction, a sense of paramount importance in many vertebrates. However there has been a progressive trend towards expansion of the telencephalon and a shift of association centers into the forebrain. The entire superficial area of the cerebral

hemispheres is an innovation (culminating in mammals) and is termed the **neopallium;** it is the major integrating area of the brain. A major evolutionary trend has been a shift in the positions of the white and gray matter. In anamniotes the gray matter is primarily internal, in keeping with the general pattern of the central nervous system; in mammals the cerebral hemispheres illustrate the shift of the gray matter to an external position. This provides unlimited opportunity for the growth of the gray region of the telencephalon. The mammalian cerebral hemispheres are the seat for the complex manipulation and thought that characterize the eutherians.

DIENCEPHALON

The diencephalon and mesencephalon are not visible dorsally owing to the tremendous expansion of the telencephalon. The diencephalon is best observed in sagittal sections and on the ventral aspect of the brain. Recall that the roof of the diencephalon is the **epithalamus,** and that the cavity of the diencephalon is the third ventricle. Posterior to the ventricle is the **pineal body (epiphysis).** Its function is obscure, but it is thought to have some endocrine effect on sexual development. The ventral region of the diencephalon is the **hypothalamus.** At the anterior border of the hypothalamus is the large **optic chiasma;** just posterior to the chiasma is a rounded elevation, the **tuber cinereum.** A narrow stalk, the **infundibulum,** suspends the **hypophysis** or **pituitary gland.**[1] The posterior aspect of the hypothalamus is bounded by a pair of **mammillary bodies.** The anterior end of the diencephalon is termed the **lamina terminalis;** dorsally there are vascular folds, the **anterior choroid plexus.** Similar vascular folds are located elsewhere in the brain. An **anterior commissure** connects the two sides of the diencephalon, and the two halves join across the middle of the third ventricle as an **intermediate mass.** The lateral walls of the diencephalon are called the **thalamus;** they form a very important relay network between the cerebral hemispheres and the rest of the brain, spinal cord, and cranial nerves. Most sensory impulses (except those of olfaction) pass

through the thalamus en route to the cerebral hemispheres. The hypothalamus, with its optic chiasma and the hypophysis, is a major associative and integrating center for such visceral and autonomic functions as temperature control, sleep, breathing, eating, and water balance. Olfaction centers rest in the tuber cinereum and mammillary bodies.

MESENCEPHALON

The ventrolateral aspect of the mesencephalon is formed by the **cerebral peduncles,** which give rise to the oculomotor nerves. The dorsal aspect is characterized by four swellings, the **corpora quadrigemina** or **colliculi.** Posteriorly, the trochlear nerves arise. The cavity of the mesencephalon is termed the **cerebral aqueduct;** it connects the third and fourth ventricles. The colliculi are all that remain of the optic lobes in mammals; but they still have important auditory and visual functions.

METENCEPHALON

The dorsal aspect of the metencephalon forms the **cerebellum;** its median, wormlike **vermis** divides the two hemispheres. In sagittal section one sees a complex branching from a white mass within the cerebellum; this is termed the **arbor vitae.** The ventral aspect of the metencephalon is the **pons;** it lies just anterior to the medulla oblongata. The trigeminal nerves arise laterally from the pons. The cerebellum is of paramount importance as a center for motor coordination and equilibrium functions.

MYELENCEPHALON

The myelencephalon constitutes the posterior region of the brain or **medulla oblongata.** At the border area between the pons and the me-

---

[1] Embryologically, the pituitary is derived from distinctive regions. Rathke's pouch, a pocket of the stomodeal (mouth) ectoderm, gives rise to the adenohypophysis (anterior lobe), or *pars distalis.* Posteriorly, where the adenohypophysis comes into contact with the infundibulum it gives rise to the intermediate lobe or *pars intermedia.* A second major region of the pituitary, the neurohypophysis (posterior lobe) or *pars nervosa,* is derived from the infundibulum of the embryonic brain. The adenohypophysis secretes at least six hormones, three of which are gonadotropic hormones and function in regulating the sex glands. A small portal system from the lower part of the hypothalamus (the median eminence) to the *pars distalis* transports specific releasing factors for each of the hormones of the anterior lobe. The *pars intermedia* secretes melanophore-stimulating hormone (MSH). The *pars nervosa* produces no hormones of its own, but serves as a storage and releasing system for two hormones (oxytocin and antidiuretic hormone in humans) produced in the anterior hypothalamus and transmitted to the *pars nervosa* by axonal transport.

dulla arise several cranial nerves. Medially the abducens (VI) emerges and laterally the facial (VII) arises in close proximity to nerve VIII. Nerve VIII is termed the vestibulocochlear in mammals, reflecting its new functions. The glossopharyngeal (IX) and vagus (X) arise in series posteriorly. The spinal accessory nerve (XI) arises by a series of rootlets to supply certain of the neck and shoulder muscles. The hypoglossal (XII) arises from the ventral aspect of the medulla and innervates muscles of the tongue and neck regions. In sagittal section the roof of the medulla may be seen to be composed of the **tela choroidea,** which, with its **choroid plexus,** overlies the fourth ventricle of the brain. The anterior part of the roof of the medulla is formed by a thin fibrous layer, termed the **anterior medullary velum.**

The medulla is arranged in much the same manner as the spinal cord, of which it is a continuation. The medulla serves as a relay station between the cerebrum and the rest of the body, but many reflex activities are handled between its nuclei; among these are breathing, swallowing, and rate of heartbeat.

The Eye

Remove an eye (**Fig. 7-11**) from the specimen of *Squalus* by cutting the extrinsic eye muscles near their insertions, along with the optic nerve and any other tissue. Upon removal of the eye one can observe the **optic pedicel,** a cartilaginous structure resembling a golf tee upon which the medial wall of the eye rests. Also note the presence of an upper and lower

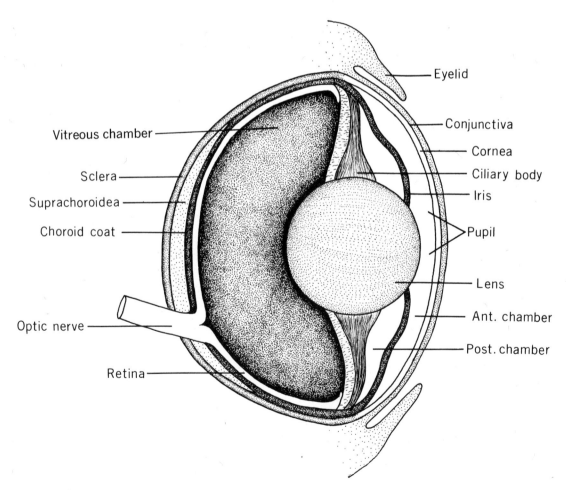

Fig. 7-11. Diagrammatic sagittal section of the eye of *Squalus*.

eyelid in *Squalus;* most fishes lack eyelids. However, in many vertebrates there is a third lid, the **nictitating membrane;** the posterior edge of this membrane can spread over the entire eye or can be retracted.

Examine the eyeball of *Squalus,* and note the tough outer covering, the **sclera** or **sclerotic coat;** it contains some cartilage, since it developed embryologically from the optic capsule of the neurocranium. Anteriorly the sclera forms the transparent **cornea,** which is covered by an epidermal layer, the **conjunctiva.** Through the cornea one can observe an opening, the **pupil,** which is surrounded by the pigmented **iris.**

Place the eyeball in a finger bowl of water and cut the eye into anterior and posterior halves. Note the crystalline **lens.** Small blackish folds form the **ciliary body** (often difficult to observe), which, along with a suspensory ligament, helps to support the lens. The large chamber behind the lens is the **vitreous chamber;** it is occupied by a fluid, the **vitreous humor.** The internal surface of the eyeball is the whitish **retina;** it contains the light receptors (the rods and cones) and is continuous with the optic nerve. Interposed between the retina and the sclera is the heavily pigmented **choroid coat.** Near the area of the optic pedicel one can observe another layer between the sclera and the choroid coat, the highly vascularized **suprachoroidea** (only present in species with a pedicel). In front of the lens is a cavity filled with a fluid; it is the **anterior chamber,** filled with the **aqueous humor.** There is a small chamber immediately posterior, between the lens and the iris; it is termed the **posterior chamber.**

The mammalian eye will not be dissected; it is essentially similar to those of *Squalus,* but differs in details. An important detail is that focusing is accomplished primarily by changing the shape of the lens; the ciliary body performs an important function in this capacity. In *Squalus* focusing is accomplished by moving the lens, rather than changing its shape.

The Ear

Remove all of the skin and associated tissue on dorsal, lateral, and posterior aspects of the neurocranium of *Squalus* to expose completely the region of the otic capsule. The position of the endolymphatic fossa will aid in your orientation. The **endolymphatic ducts** connect with a series of canals and sacs collectively known as the **membranous labyrinth** (**Fig. 7-12**); these canals are filled with a fluid known as **endolymph,** which in elasmobranchs may be composed primarily of sea water from the endolymphatic ducts. Surrounding the membranous labyrinth is a series of encompassing canals, the **osseous labyrinth** (cartilaginous in elasmobranchs), also filled with a fluid, the **perilymph,** which enters via the **perilymphatic ducts.**

Carefully begin scraping away the cartilage covering the otic region; soon you will begin to expose the **anterior** and **posterior semicircular canals.** As you proceed you will expose the **horizontal canal,** and the **sacculus** and **utriculus.** Within the sacculus and utriculus are small calcareous granules; calcium salts may combine with sand grains that have entered via the invagination canals to form a mass known as the **otolith.** Note that on each canal there is a slight swelling; these are termed the **ampullae,** and contain endings of the statoacoustic nerve (VIII). On the posteroventral surface of the sacullus is a bulge known as the **lagena.** Remove the entire inner ear and place it in a bowl of water for study.

The inner ear of fishes is primarily an organ of equilibrium, though there is some evidence that in some fishes it also performs an auditory function. In tetrapods the inner ear continues to serve an equilibrium function, but its structure is further elaborated. Recall that in all tetrapods there is a middle ear in addition to the inner ear (see Fig. 4-15). Sound waves impinge on the ear drum or tympanic membrane, and are transmitted through the middle ear cavity

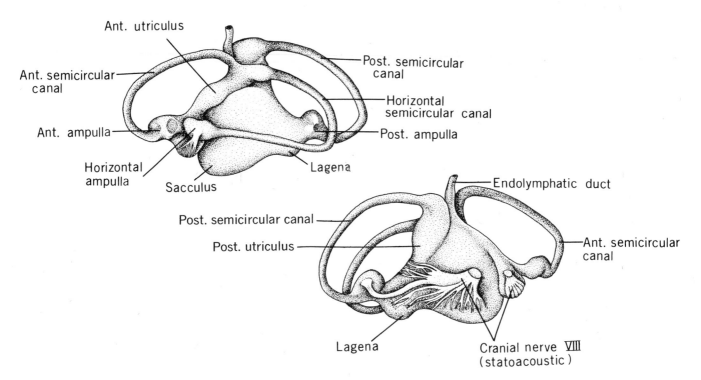

Fig. 7-12. Lateral *(above)* and medial *(below)* views of the membranous labyrinth of *Squalus* (left ear). (Drawn from specimens and following Marinelli and Strenger.)

via the stapes. Recall that the middle ear cavity is homologous to the old spiracular cavity, and the stapes is homologous to the fish hyomandibular. The stapes has a flat footplate that inserts into an oval window of the inner ear (on the side of the otic capsule), and thus sound vibrations are transformed into fluid waves in the perilymph of the inner ear. The perilymph transmits these waves to the part of the membranous labyrinth responsible for transforming them into nerve impulses. The round window is responsible for releasing these vibrations. In the lower tetrapods, it is thought, the lagena is the part of the labyrinth responsible for the auditory function; in mammals a cochlea (see Figs. 4-15 and **7-13**) has developed as a complex coiled structure for sound detection, and two additional bones, the malleus and incus, form two of the three links in the ossicular chain. In addition, a pinna may develop, partially surrounding the external auditory meatus (external ear), to aid in the initial reception of sound waves.

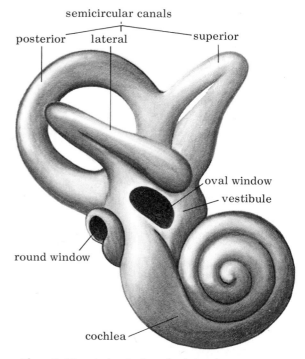

Fig. 7-13. Labyrinth of right human ear. (From Keeton, Fig. 10.40, modified from Sobotta-Figge: *Atlas of Human Anatomy,* 8th English ed., Hafner Publishing Co., New York.)

8

Structures of the Coelom and Pharynx

A **coelom** is a body cavity lined by a mesodermally derived epithelium. When the embryonic hypomere or lateral plate splits, it leaves bilateral coelomic spaces, with mesenteries suspending the abdomnial viscera by **dorsal** and **ventral mesenteries** (see Fig. 6-1). The coelomic spaces are divided very early in evolution into **pericardial** and **pleuroperitoneal** cavities (**Fig. 8-1B**); a **transverse septum** separates the two cavities. In the amphibians another pair of coelomic cavities develops dorsal to the pericardial cavity to house the lungs; these are the **pleural cavities,** and they are partitioned by mesenteric folds from the dorsal mesentery and body wall. The new membrane is typically called the **pleuroperitoneal membrane.** In mammals, the tissues of the transverse septum and the pleuroperitoneal membranes combine as an oblique septum across the coelomic cavity; these membranes are invaded by cervico-occipital myotomes, and innervated by nerves from the same region; the resulting partition is known as the **diaphragm,** and the posterior compartment is the **peritoneal cavity** (Fig. 8-1C).

The pharynx begins its embryological development as a solid-walled structure. Soon, however, furrows develop in the outer ectoderm and the inner endoderm; when the furrows meet, the visceral pouches are formed. The tissues between these pouches are the visceral arches, containing branchiomeric muscle, cranial nerves, aortic arches, and visceral cartilages. In fishes such as *Squalus* the spiracle is formed from the first pouch, but primitively it was a fully developed gill arch. In higher vertebrates the middle ear cavity and eustachian tube are developed in the first pouch. In mammals the **palatine tonsils** develop at the site of the second pouch, and in all vertebrates a **thymus** develops in the area of the posterior visceral arches. Thymosin, secreted by the thymus, functions to stimulate immunologic competence in lymphoid tissues and plasma cells.

A **thyroid gland,** probably homologous to the endostyle of the protochordates, develops in the floor of the pharynx in the region of the first and second pouches. Most vertebrates have a pair of thyroid glands, but in man they are fused into a single structure (see Fig. 8-8). The principle thyroid hormones, thyroxin and triiodothyronine, function in some capacity to speed up oxidative metabolism. The synthesis of thyroid hormones is stimulated by thyroid-stimulating hormone from the adenohypophysis. The thyroid gland also produces a third hormone, calcitonin, which prevents an

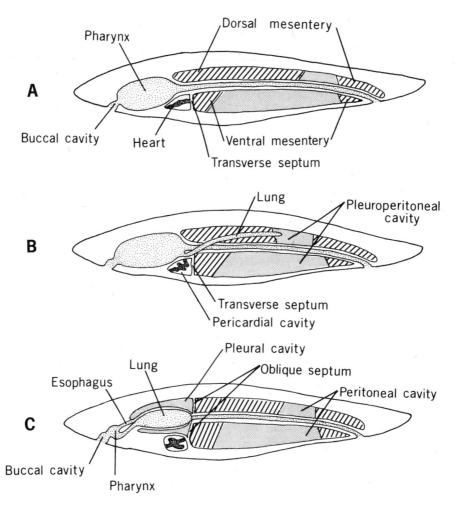

Fig. 8-1. Diagrammatic illustrations showing the relationships of the different parts of the coelom in a fish (*A*), an amphibian (*B*), and an advanced reptile (*C*). (After Smith.)

excessive rise of calcium in the blood by acting antagonistically to the parathyroid hormone (parathormone). The **parathyroid glands** (not present in fishes) are small glands (usually four) located on the surface of, or near, the thyroid gland. Parathormone functions in the regulation of calcium concentration between the blood and other tissues, particularly bone. A rise in parathormone levels causes an increase in the concentration of calcium and a decrease in the concentration of phosphate in the blood by acting to prevent the kidneys and intestine from excreting calcium, and acting to stimulate the release of calcium from bone.

Ultimobranchial bodies develop in all verte-brates in the area of the last pair of visceral arches. In most vertebrates the ultimobranchial bodies probably synthesize calcitonin; however, in mammals the embryonic tissue of the bodies contributes to glandular tissue in the area of the parathyroid and thyroid glands.

Two **adrenal glands** lie within the body cavity near the cephalic end of the kidneys in all vertebrates. In mammals these glands have distinctive cortical and medullary regions; in many vertebrates, however, the distinction is less apparent. The **adrenal medulla** is an ectodermal derivative, being derived embryologically from the neural crest cells. It synthesizes catecholamines, principally epinephrine or adrenaline, and norepinephrine or noradrenalin. The post-ganglionic neurons of the sympathetic ganglia are homologous to the cells of the adrenal medulla. Both are derived from neural crests, and both are stimulated to secrete catecholamines by preganglionic neurons.

The **adrenal cortex** is derived from the same mesodermal tissue that gives rise to the gonads. There is a very large number of cortical hormones, but all are steroid substances, chemically very similar. These hormones act in regulating salt and water balance, in regulating carbohydrate and protein metabolism, and also as sex hormones.

Immediately posterior to the last visceral pouch, **lungs** may develop as evaginations of the floor of the pharynx. Lungs are very primitive structures and are thought to have occurred in very early fishes. In advanced fishes they have become transformed into hydrostatic organs, the **swim bladders.**

The gut itself, which is formed from the embryonic archenteron, is an endodermally lined structure that has some ectodermal contributions from anterior (**stomodeal**) and posterior (**proctodeal**) invaginations; part of the pituitary is of stomodeal origin. One may divide the archenteron into fore- and hindguts; the former includes the **pharynx, esophagus,** and **stomach,** the latter, the **intestinal region** and **cloaca.** At the level of transition between the fore- and hindguts, the **liver** and **pancreas** arise, the liver as a ventral diverticulum of the archenteron. The liver is a very important organ, serving predominantly in gross metabolic activities and conversions, and also in the secretion of bile. The pancreas is usually both a dorsal and ventral evagination. The exocrine portion of the pancreas secretes digestive enzymes; the endocrine islets of Langerhans produce insulin, a substance necessary for carbohydrate metabolism. Posteriorly, a **urinary bladder** is present in tetrapods, but not in fishes. In mammals the cloaca is divided into the dorsal **rectum** and the ventral **urogenital sinus.**

Fishes

The pharynx and coelom of *Squalus* exhibit many features of the typical primitive fishes, except that in the primitive condition (1) the first gill pouch was not modified as a spiracle, (2) a stomach was probably not definable, and (3) lungs were often present.

In order to observe the pharynx and coelom, an incision should be made through the angle of the mouth, cutting through the middle of the external gill slits, visceral arches, and the pectoral girdle. Extend the incision medially and to the other side (as in **Fig. 8-2**). In addition, make a longitudinal incision from the pectoral girdle to the cloaca. Transverse incisions may be made about midway along the body wall to aid in exposing the organs.

Within the **pharynx** (Fig. 8-2) there is an immovable **tongue,** which is supported by the hyoid arch. In normal activity water enters the mouth and **spiracle,** and is shunted out the external gill slits. Numerous **gill rakers** project from each internal arch; these prevent food particles from entering the gill chambers. The pharynx continues into the **esophagus,** which bears numerous papillae.

The digestive organs lie within the pleuroperitoneal cavity and are suspended by a mesentery whose surfaces constitute the **visceral peritoneum.** The body wall is lined with the **parietal peritoneum.** The heart lies within another coelomic cavity, the pericardial cavity, which is located anterior to the pectoral girdle. This cavity is surrounded by the **parietal pericardium,** the heart itself by the **visceral pericardium.** A transverse septum separates the pericardial from the pleuroperitoneal cavity.

The esophagus is continuous with the J-shaped **stomach,** which terminates in a short arm, the **pyloric portion** (**Fig. 8-3**). This portion terminates in the **pyloric sphincter.** This sphincter marks the beginning of the small in-

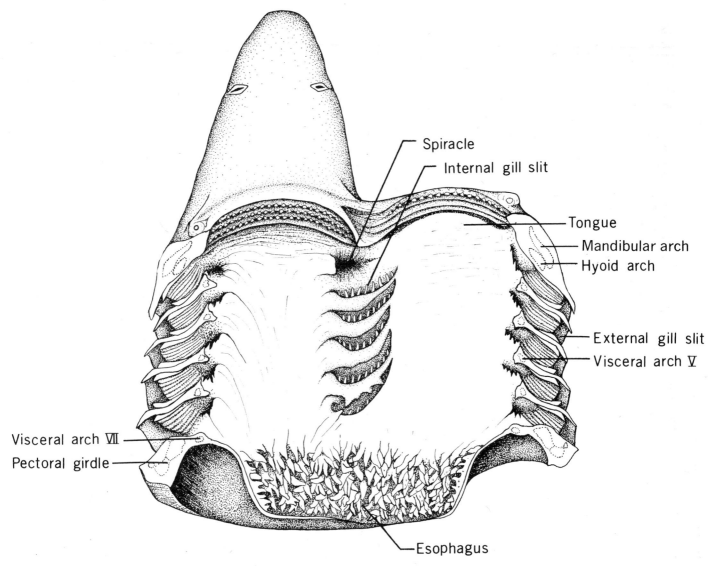

Fig. 8-2. The pharyngeal region of *Squalus*.

testine. The anteriormost portion of the small intestine is known as the **duodenum;** the posterior portion is the **spiral valve.**

Cut open the spiral valve to observe its structure. Its spiral staircase structure serves to slow down food to ensure complete digestion, and also serves to increase the internal area of the intestinal tract. Many of the primitive bony fishes had spiral valves. The remaining short portion of the intestine is the **large intestine;** it terminates in the anterior wall of the **cloaca.** The small digitiform (finger-shaped) **rectal gland** extends dorsally from the rectum; it is a salt-excreting gland. Its dorsal mesentery is

known as the **mesorectum** or **mesocolon,** which also suspends the large intestine. Note that the esophagus and stomach are suspended by a large dorsal mesentery called the **greater omentum** or **mesogaster;** the mesentery of the small intestine is simply called the **mesentery.** The **spleen** lies within the greater omentum, but its mesentery is given the special name **gastrosplenic ligament,** since it is attached to the greater curvature of the stomach.

The **liver** consists of two very large right and left lobes and a small median lobe that contains the **gall bladder.** The gall bladder is easily recognized by its greenish color. The **falciform**

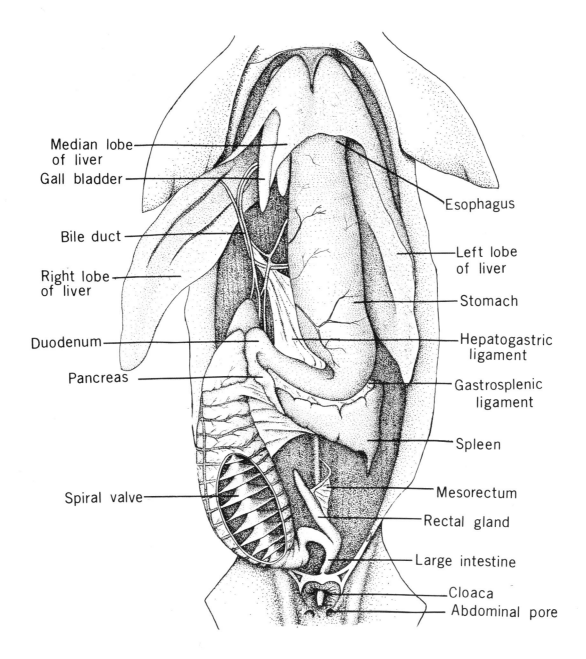

Median lobe
of liver

Gall bladder

Bile duct

Right lobe
of liver

Duodenum

Pancreas

Spiral valve

Esophagus

Left lobe
of liver

Stomach

Hepatogastric
ligament

Gastrosplenic
ligament

Spleen

Mesorectum

Rectal gland

Large intestine

Cloaca

Abdominal pore

Fig. 8-3.   Ventral view of the abdominal viscera of *Squalus.*

ligament connects the anterior end of the liver with the midventral body wall. The falciform ligament is all that remains of the ventral mesenteries of the anamniotes. The liver is connected to the posterior ends of the stomach and the duodenum by a broad mesentery, the **lesser omentum.** The lesser omentum actually consists of two converging mesenteries, one from the stomach, the **hepatogastric ligament,** and one from the duodenum, the **hepatoduodenal liga-** ment; the latter carries the bile duct to the duodenum. Three major structures are within the lesser omentum: (1) the hepatic artery, (2) the hepatic portal vein, and (3) the bile duct (ductus choledochus). The **bile duct** passes from the liver to the lumen of the duodenum. The **pancreas,** which consists of dorsal and ventral lobes, has a duct that empties into the intestine in the same area as the bile duct, but the duct is difficult to observe.

Primitive Tetrapods

In order to observe the pharynx and coelom (**Figs. 8-4** and **8-5**), incisions should be made in *Necturus* equivalent to those made in *Squalus*.

Because *Necturus* is neotenic, it is in a state of suspended development between aquatic and terrestrial existence. Although lungs are present, the major breathing organs are the three **external gills** with their branching gill filaments. There are only two gill slits; they may be found by parting the external gills. A tongue is present, but it is developed little more than that of *Squalus*.

Necturus can also breathe atmospheric oxygen. Air is taken in by the **external nares** and enters the pharyngeal cavity via the **choanae** or internal nares. Air enters the actual respiratory tract through the **glottis**; from there it passes into the **laryngotracheal chamber** to the

bronchi and **lungs.** One can determine these relationships by passing a probe through a cut in the lungs and out the glottis.

The pharynx continues posteriorly as the **esophagus,** which becomes enlarged posteriorly as the **stomach.** As in *Squalus,* the stomach terminates as the **pylorus,** and the **small intestine** begins as the **duodenum.** The **pancreas** is present in the area of the duodenum. The small intestine continues, making several convolutions, and shortly before reaching the **cloaca,** the intestine enlarges to become the **large intestine.** The **urinary bladder** is present ventral to the large intestine. As in *Squalus,* the **mesogaster** (**greater omentum**) suspends the stomach. There is a **gastrosplenic ligament,** which suspends the **spleen** (an elongate, digitiform organ) from the stomach. Mesenteries equivalent to the hepatic ligaments of *Squalus* are also present. The large **liver** (which is somewhat displaced toward the right side of the coelom) is not divided into distinctive lobes as in

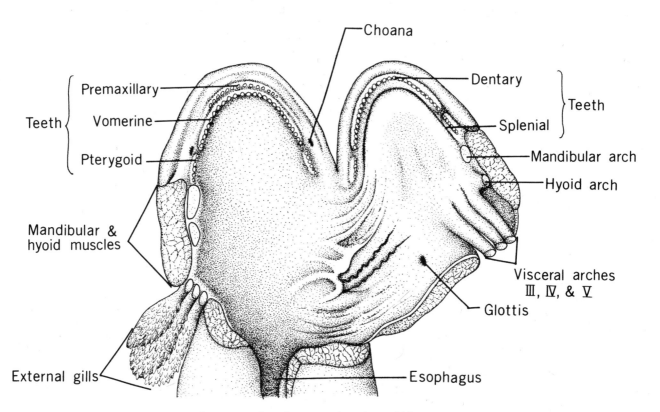

Fig. 8-4. The pharyngeal region of *Necturus*.

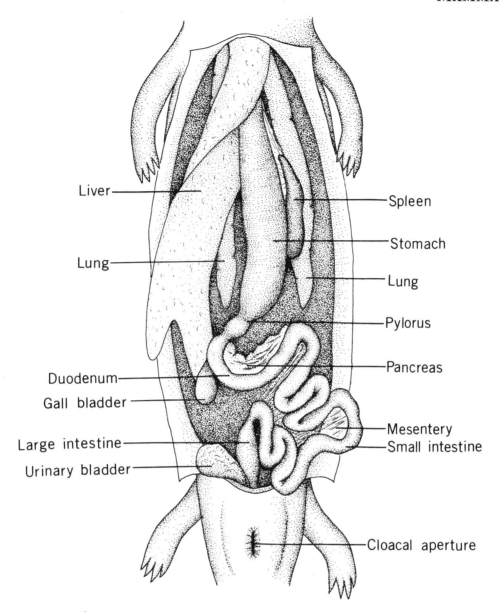

Liver

Spleen

Stomach

Lung

Lung

Pylorus

Pancreas

Duodenum

Gall bladder

Mesentery

Large intestine

Small intestine

Urinary bladder

Cloacal aperture

Fig. 8-5. Ventral view of the abdominal viscera of *Necturus*.

Squalus. The **gall bladder** is present near the posterior end of the liver, and sends the **bile duct** to the duodenum.

As in *Squalus*, the **heart** occupies an area anterior to the pectoral girdle, beneath the hypobranchial musculature. The pericardial cavity is lined with the **parietal pericardium**, and the heart is covered with the **visceral pericardium.** A vertical **transverse septum** separates the pericardial and pleuroperitoneal cavities, but the lungs are still within the latter, and do not have a separate coelomic division.

Mammals

The lower tetrapods are all ectothermic, deriving their major source of body heat from outside the body, but mammals and birds are endothermic—their major source of heat is within the body from muscle activity and other means of thermogenesis. Having this constant source of heat allows the organism to maintain a constant metabolic rate, and to remain active throughout the day. Many organ systems have become modified for the endothermic mode of life. A highly movable head and neck region have also evolved in relation to this greatly increased activity.

To reveal the structures of the mouth and pharynx it is necessary to make a sagittal section of the head and neck, and remove the skin from one side; this may already have been done in your studies of the CNS.

Salivary glands have evolved from small oral glands in the primitive tetrapods. Secretions from these glands lubricate the food and break down carbohydrates by the action of the enzyme ptyalin. Beneath the auricle of the ear is the large **parotid** gland (**Fig. 8-6**); its duct crosses the masseter muscle and enters the mucous membrane of the upper lip. **Mandibular** and **sublingual glands** are present ventrally, along with numerous lymph nodes.

The **tongue** is a very important organ in mammals for manipulating the food. It has become invaded by a large mass of hypobranchial muscle, which is innervated by motor fibers of the hypoglossal nerve. In addition, lateral lingual swellings have contributed to the tongue. Sensory innervation is as one would predict from previous studies of the cranial nerves. The trigeminal (V) innervates the anterior portion, which is derived from lingual swellings in the mandibular arch area; posteriorly, nerves VII and IX take over. The tongue is connected to the floor of the mouth by the **lingual frenulum.** The dorsal surface of the tongue is covered with **papillae.**

The portion of the mouth between the lips and the teeth is commonly known as the **vestibule;** the portion behind the teeth, the **oral cavity** (**Fig. 8-7**). The **secondary palate,** composed of the **hard palate** anteriorly and the fleshy **soft palate** posteriorly, separates the oral cavity and anterior pharynx from the nasopharynx, thereby allowing breathing to occur simultaneously with food manipulation within the oral cavity. Posteriorly, food is prevented from entering the larynx by the **epiglottis,** which guards its opening, the **glottis.** The res-

piratory passages have become highly modified in comparison with their state in primitive tetrapods. The original laryngotracheal chamber is now subdivided into the anterior **larynx,** which also houses the vocal cords, and the posterior **trachea,** which descends caudally and splits into two bronchi before entering the lungs. Movement of air across the vocal cords sets them in motion.

Glossopalatine arches are the lateral folds that extend from the posterior region of the tongue to the soft palate. These folds separate the oral cavity from the pharynx proper. A pair of **palatine tonsils** (masses of lymphoid tissue) are located in the lateral walls of the posterior oral cavity. In the posterolateral walls of the nasopharynx are located the openings of the **eustachian tubes,** which communicate with the middle ear cavity. Eustachian tubes are derived from the first visceral pouch.

One may open one of the thoracic cavities by making an incision slightly to one side of the midventral line and extending the entire length of the sternum; strong scissors are necessary to make such a cut. Upon opening the incision, one of the pleural cavities will be revealed. These cavities are lined with a mesentery, the **parietal pleura;** the lungs are covered with **visceral pleura.** The two pleural cavities are separated by a space, the **mediastinum,** which is largely occupied by the pericardial cavity, but which exhibits ventrally a mediastinal septum.

The **trachea** branches into paired **bronchi** before entering the lungs. The **lungs** are complex, multi-lobed structures with labyrinthine passageways terminating in small pockets, the **alveoli,** in which actual gas exchange occurs. Mammalian lungs are greatly elaborated in comparison with the lungs of primitive tetrapods, and the actual surface area for gas exchange has increased tremendously. In addition, cervico-occipital myotomes have invaded the partition of the pleural and peritoneal cavities, giving rise to the muscular **diaphragm,** which, along with rib movement, provides a mechanism for ventilating the lungs.

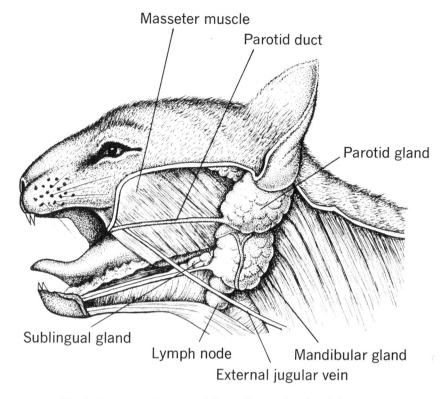

Fig. 8-6. Lateral view of the salivary glands of the cat.

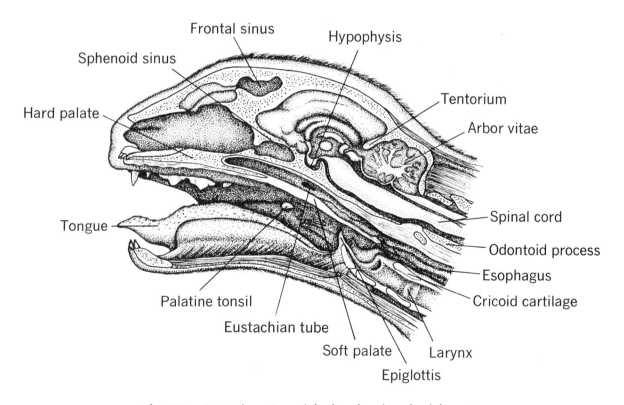

Fig. 8-7. Sagittal section of the head and neck of the cat.

A pair of phrenic nerves (located on either side of the mediastinum) innervates the diaphragm,

The darkly colored **thyroid gland** may be observed at the anterior end of the trachea. There are two pairs of **parathyroid glands** embedded in the dorsal region of the thyroid (**Fig. 8-8**), but they cannot be seen. The **thymus gland** is located in the mediastinum, ventral and anterior to the heart.

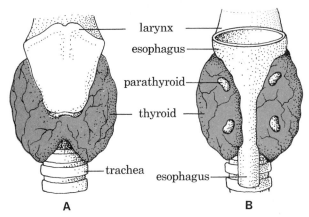

Fig. 8-8. Human thyroid and parathyroid glands. *A.* Ventral view of thyroid. *B.* Dorsal view of thyroid, with parathyroids. (From Keeton, Fig. 9.20.)

The heart and pericardial cavity occupy most of the area of the mediastinum (**Fig. 8-9**). The pericardial cavity may be revealed by making an incision slightly to one side of the midventral line from the diaphragm to the pelvic girdle. Transverse incisions may also be useful in obtaining access to the coelom. The coelom (as in all vertebrates) is lined by the **parietal peritoneum;** the viscera are suspended by the **visceral peritoneum.**

The large **liver** fits into the concave surface of the diaghragm; the **falciform ligament** connects the liver, diaphragm, and ventral body wall. The free end of this ligament represents the vestige of the umbilical vein and is called the **round ligament** of the liver. **Coronary ligaments** are present dorsally between the diaphragm and the liver. The liver is divided into right and left lobes, which are further subdivided into medial and lateral lobes. In addition, there is a caudate lobe, which is present posterior to the right lateral lobe and is in close proximity to the kidney. The **gall bladder** lies in the right medial lobe of the liver. Two **hepatic ducts** from each side of the liver join the **cystic duct** from the gall bladder to form the **common bile duct** which empties into the duodenum near the pyloric sphincter.

The **esophagus** is a narrow tube that extends from the pharynx to the stomach. It lies dorsal to the larynx and trachea, passes through the mediastinum, pierces the diaphragm, and continues in the peritoneal cavity to the dilated stomach, which it joins at the cardiac end. The large **stomach** has a caudal pyloric region that courses anteriorly and terminates at the **pyloric sphincter;** the digestive tube then continues as the **duodenum.** The flattened elongate body lying in the area between the duodenum and pyloric stomach is the **pancreas.** The **pancreatic duct** may be observed by picking away the pancreatic tissue. It joins the common bile duct before emptying into the duodenum (approximately 3 cm from the pyloric sphincter). As in the primitive tetrapods, the pancreas has both endocrine and exocrine functions.

The **lesser omentum** is the part of the ventral mesentery that extends from the liver to the stomach. It is divided into a **hepatogastric ligament,** extending to the stomach, and a **hepatoduodenal ligament,** extending to the duodenum. The large **greater omentum** or **mesogaster** is part of the dorsal mesentery, which attaches along the greater curvature of the stomach and forms the saclike **omental bursa** that drapes down over the intestines. The **spleen** is a rather large organ in the cat, and lies in the omentum on the left side of the stomach. The **rugae** of the stomach may be revealed by opening the stomach with a scalpel.

The **small intestine** consists of several divisions. The duodenum has already been discussed; it is the first U-shaped segment of the small intestine, which then continues as the **jejunum.** The posterior half of the small intestine is known as the **ileum,** but there is no clear-cut distinction between the jejunum and

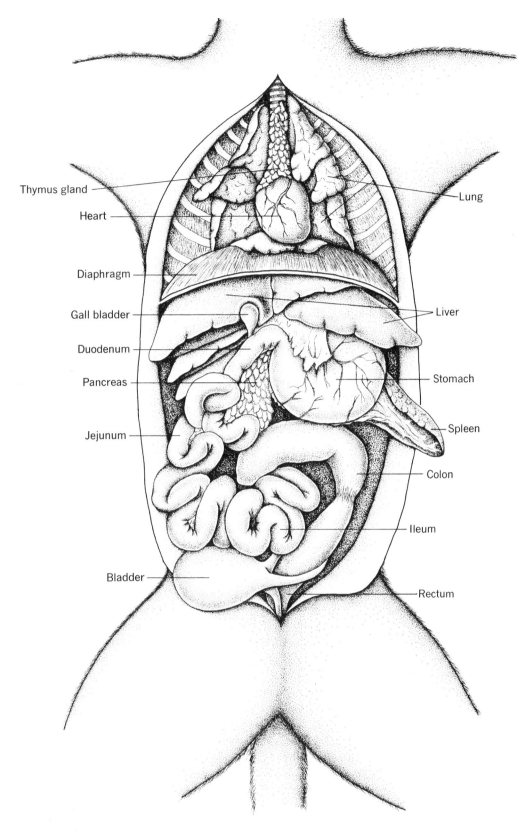

Fig. 8-9. Ventral view of the thoracic and abdominal viscera of the cat.

the ileum. The small intestine is suspended by a portion of the dorsal mesentery called the **mesentery.** The large intestine is longer than in the primitive tetrapods. Slightly posterior to the point of union of the ileum and large intestine is a blind pouch, the **caecum,** which is somewhat comparable to the human appendix. It is greatly increased in herbivores, and contains bacteria that aid in the digestion of cellulose. Most of the large intestine is called the **colon** and is suspended by the **mesocolon.** The colon finally continues into its terminal segment, known as the **rectum,** which opens to the exterior via the **anus.**

9

The Circulatory System

The circulatory system of vertebrates carries the blood to and from the various parts of the body. It consists of the arteries, veins, capillaries, and the muscular pumping organ, the heart. Lymph is carried in a second part of the circulatory system, the lymphatic system, which consists of the lymphatic vessels and lymph nodes. Blood exchanges carbon dioxide for oxygen in the respiratory system, and carries oxygen, along with nutrients, hormones, and so forth, to the tissues of the body. It transports excretory products to the appropriate excretory organs. In addition, the circulatory system functions in protecting the organism from disease, and in transporting heat for thermoregulation. The lymphatic system takes up interstitial fluids not collected by the venules into its channels and returns them to the venous system; it also transports emulsified fats from the intestine to the venous system. The spleen was discussed with the organs of the coelom; it is somewhat like a lymph node, but usually serves as a site for the storage and production of blood cells.

Blood leaving the heart travels through the muscular arteries. Arteries are elastic structures capable of considerable distension. Arteries terminate in capillary beds, where considerable fluid leaves the vessels to circulate within the tissues. At the venule end of the capillaries, the interstitial fluid flows back into the capillaries and is transported back to the heart via the veins. Veins are thin-walled structures with valves that prevent the backflow of blood. The lymphatic vessels pick up the fluid not retrieved by the venous system, which may be considerable. Lymphatic channels, like the veins, have valves preventing backflow. Fishes, amphibians, and reptiles have lymph hearts to pump lymph back to the venous channels. However, endotherms lack these structures and the lymph is propelled through the lymphatics by respiratory, muscular, intestinal, and other movements. The fluid, or hemolymph, is emptied into the venous channels via several thoracic ducts in the area of the jugular veins (see Fig. 9-11).

The heart (**Fig. 9-1**) is a large midventral muscular pumping organ that may be thought of as a modified blood vessel. In fishes it pumps unoxygenated blood forward ventrally through the ventral aorta, which gives rise to a series of afferent branchial arteries to the various gills. These terminate in capillary beds in the gill lamellae, where gas exchange takes place. Oxygenated blood then courses dorsally through a series of efferent branchial arteries, which join dorsally to form a dorsal aorta. Thus each visceral arch contains a blood vessel that connects the ventral and dorsal aortae. (This holds true for all vertebrates, at least in the embryonic stage.) These blood vessels are known as the aortic arches; their evolutionary history is diagrammed in **Fig. 9-2.** In all vertebrates six pairs of aortic arches develop, at least transitorily. Typically there is a pair of aortic arches in front of each visceral pouch; thus the first aortic arch is housed in the first visceral arch.

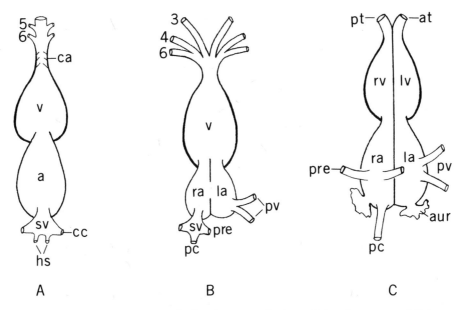

Fig. 9-1.   Diagrammatic illustrations of the hearts of: *A*, a fish; *B*, an amphibian; and *C*, a mammal. The structures shown are: hs, hepatic sinus; cc, common cardinal vein; sv, sinus venosus; a, atrium; v, ventricle; ca, conus arteriosus; 5 & 6, fifth and sixth aortic arches; pc, postcava; pre, precava; ra, right atrium; la, left atrium; pv, pulmonary veins; 3, 4 & 6, third, fourth, and sixth aortic arches; aur, auricular flap; rv, right ventricle; lv, left ventricle; pt, pulmonary trunk; at, aortic trunk. (Slightly modified after Kent, *Comparative anatomy of the vertebrates,* 3rd ed. Courtesy The C. V. Mosby Company, St. Louis.)

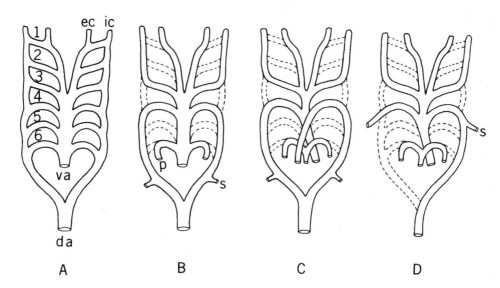

Fig. 9-2.   Diagrammatic illustration of the pattern of aortic arches in: *A*, a fish (basic pattern); *B*, an amphibian; *C*, a reptile; and *D*, a mammal. Shown are: da, dorsal aorta; ec, external carotid; ic, internal carotid; p, pulmonary artery; s, subclavian artery; va, ventral aorta.

Fishes

The circulatory system of *Squalus* aptly illustrates the primitive piscine pattern. The pericardial cavity has been described previously. The heart in all fishes is similarly constructed. It consists of four chambers; from posterior to anterior, they are the sinus venosus, atrium, ventricle, and conus or truncus arteriosus (**Fig. 9-3**). The **sinus venosus** is a very thin-walled, triangular structure that is in contact with the transverse septum. It collects unoxygenated blood from the liver via the **hepatic sinuses,** and from the various regions of the body via the **common cardinal veins (ducts of Cuvier)** and passes this blood into the **atrium,** which lies anterior to the sinus venosus and dorsal to the ventricle. Blood enters the **ventricle** from the atrium. The ventricle is a thick-walled, muscular chamber; it is the pumping organ of the piscine heart. On the ventral surface are the coronary arteries. Blood pumped out of the ventricle enters the **conus arteriosus,** a straight muscular tube that is continuous anteriorly with the ventral aorta.

Reveal the **ventral aorta** along its entire length by removing the hypobranchial musculature, and trace the afferent branchial arteries into the gill arches. Five pairs of **afferent branchial arteries** extend up into the visceral arches. At the anterior bifurcation of the ventral aorta, the **dark thyroid gland** may be observed anteriorly. The oxygenated blood is received from the gill lamellae by a series of **efferent branchial arteries.** Each has its beginning as a **collector loop** formed of **pre-** and **posttrematic** branches. Cross trunks pass through the interbranchial septa. The last gill has an incomplete set of these arteries.

The efferent branchial arteries may be seen emerging from the gills by removing the membrane from the roof of the oral cavity and pharynx (**Fig. 9-4**). There are four pairs of efferent branchial arteries; they converge dorsomedially to form the large **dorsal aorta.** Though most of the blood coursing posteriorly to the body travels via the dorsal aorta, a small **pharyngo-esophageal artery** that arises from the

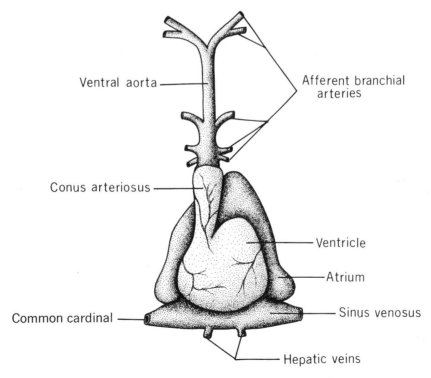

Fig. 9-3. Ventral view of the heart of *Squalus*.

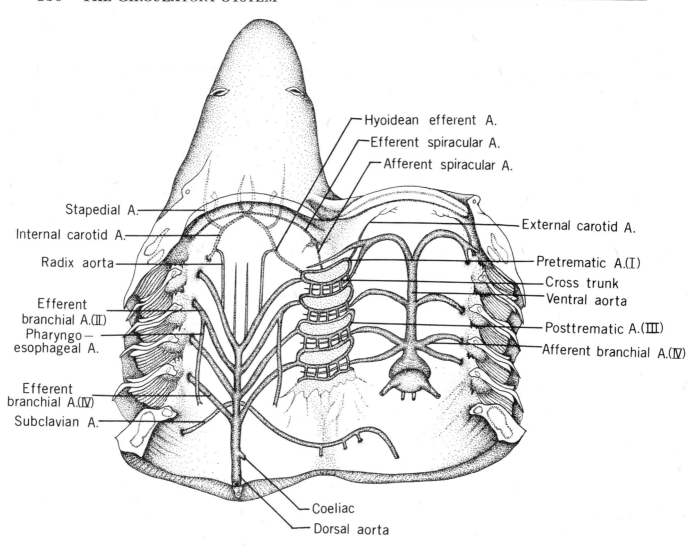

Fig. 9-4. Afferent and efferent branchial arteries and associated arteries of *Squalus*.

second efferent branchial artery supplies the roof of the pharynx and the esophagus. A pair of **subclavian arteries** arises between the third and fourth efferent branchial arteries and later gives rise to a number of minor vessels extending into the pectoral fin. Eventually these become the **brachial arteries,** located distally in the fin.

Several other channels exist for carrying oxygenated blood to the anterior parts of the body. Arising from the dorsal end of the first collector loop on each side is the **hyoidean efferent artery.** The hyoidean efferents are joined about halfway along their length by paired **radix aortae;** these represent the anterior remnants of the originally paired dorsal aortae. After the

point of joining of the radix aortae, the hyoidean efferents continue as the **internal carotids.** The internal carotids course anteriorly, enter the neurocranium via the carotid foramen after uniting, and are the main arteries supplying the brain. Midway along the course to the carotid foramen, the internal carotids give rise to the **stapedials,** which go to structures within the orbit. A small vessel, the **afferent spiracular artery,** arises from the pretrematic of the first collector loop; it goes to the pseudobranch of the spiracle. An **efferent spiracular** extends medially from the pseudobranch to connect with the internal carotid. From the ventral end of the first collector loop arises an **external carotid artery;** it supplies the region of the lower jaw.

DORSAL AORTA AND BRANCHES

In *Squalus* (**Fig. 9-5**) many branches arise from the dorsal aorta. Between body segments, paired intersegmental arteries send dorsal rami to the epaxial region, and ventral rami to the hypaxial region. In addition, small, paired, lateral visceral arteries send numerous branches to the kidneys and gonads. Other small, ventral, visceral branches pass into the mesenteries and viscera, etc. This basic pattern is fairly constant in vertebrates, and even the larger branches of the dorsal aorta are modifications on this theme. The major branches of the dorsal aorta are listed in sequence, from anterior to posterior:

Subclavians (modified intersegmental): These paired arteries arise from the aorta between the entrance of the third and fourth efferent

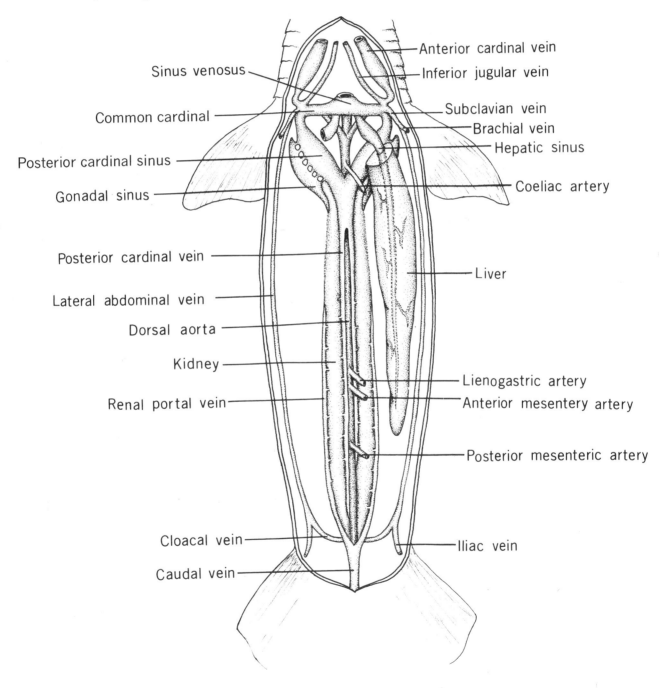

Fig. 9-5. The circulatory system of *Squalus*.

branchial arteries. They pass into the pectoral appendage.

Coeliac (modified ventral visceral): The coeliac is the most anterior unpaired branch of the dorsal aorta in the peritoneal cavity. The coeliac gives off small branches to the esophagus, stomach, and gonads, then continues to the gastrohepatic ligament, where it divides to give rise to:

(1) the **pancreaticomesenteric,** leading to the pancreas, pylorus, duodenum, and, as the anterior intestinal, to the anterior region of the spiral valve, and

(2) the **gastrohepatic,** which divides into the gastric, leading to the stomach, and the hepatic, leading to the liver.

Lienogastric (modified ventral visceral): This artery passes to the spleen and the posterior region of the stomach.

Anterior mesenteric (modified ventral visceral): The anterior mesenteric passes to the posterior part of the spiral valve.

Posterior mesenteric (modified ventral visceral): This artery goes to the rectal gland and the posterior region of the intestine.

Iliacs (modified intersegmentals): The iliacs enter the pelvic fins.

Caudal: The caudal artery is the continuation of the dorsal aorta into the tail region.

SYSTEMIC VEINS

The systemic veins of vertebrates are those that empty directly into the posterior regions of the heart. *Squalus* exhibits (with some modifications) most of the primitive features of the systemic veins. However, veins in elsmobranchs are very much like sinuses; this is correlated with the elasmobranchs' low blood pressure and high blood volume as compared with the other fishes. The systemic veins are, from anterior to posterior:

Hepatic sinuses: These veins carry venous blood from the lateral lobes of the liver to the sinus venosus.

Inferior jugulars: These veins drain the floor of the branchial region, and empty into the anterior wall of the common cardinals.

Common cardinals (**ducts of Cuvier**): These are the major collecting channels that empty into the lateral aspects of the sinus venosus.

Lateral abdominals: The lateral abdominal veins drain the posterior appendages and the cloacal regions (by the femoral and cloacal veins), continue in the lateral abdominal wall, and anteriorly join the **brachials** (which drain the pelvic appendages) to form the **subclavian,** which enters the common cardinals.

Anterior cardinals: The anterior cardinals drain the brain and most of the head, and empty into the sinus venosus.

Posterior cardinals: The posterior cardinals drain the dorsal body wall and the kidneys, expand as the posterior cardinal sinus, and empty into the common cardinals.

THE HEPATIC PORTAL SYSTEM

Portal systems are venous channels that begin in the capillary beds of a region and terminate in the capillary beds of another region. The hepatic portal system is invariably present and terminates in the capillary beds of the liver. In triple-injected specimens, the hepatic portal system is injected with yellow latex.

Hepatic portal vein: This is the major vein of the system. It lies along the bile duct in the gastrohepatoduodenal ligament, and receives three major veins. The **gastric,** from the cardiac region of the stomach, the **lienomesenteric,** from the pancreas, spleen, and part of the spiral valve, and the **gastrointestinal,** from the pylorus, spleen, part of the spiral valve, all merge into the hepatic portal vein.

THE RENAL PORTAL SYSTEM

The renal portal system drains the caudal region of the body and terminates in the capillary beds of the kidneys. The single **caudal vein** bifurcates at the level of the cloaca to form the paired **renal portal veins,** located dorsal to the opisthonephroi.

Primitive Tetrapods

The transition to land involved a switch from branchial to pulmonary respiration; this change is reflected primarily in the region of the heart and aortic arches. The heart tends to become laterally partitioned. Blood from the body tends to enter the right side of the heart; blood from the lungs, the left side (see Fig. 9-1). In most amphibians only the atria are separated, and some mixing of oxygenated and unoxy-genated blood may occur in the ventricle. In reptiles there is a partial separation of the ventricles, and in the crocodilians the ventricular septum is complete.

Necturus illustrates most of the features of the circulatory pattern of primitive tetrapods (**Fig. 9-6**), but being neotenic, it still relies on larval circulation. The circulatory system is difficult to dissect, and the major features should therefore be shown through demonstration.

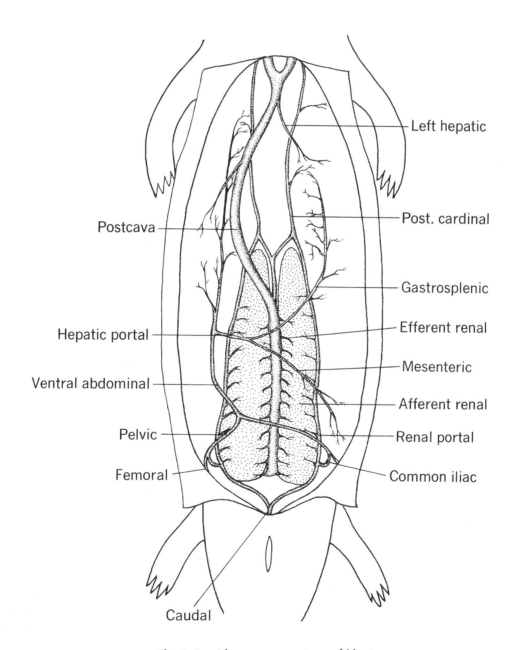

Fig. 9-6. The venous system of *Necturus*.

Necturus retains the piscine sinus venosus, but the atrium has become divided into separate right and left halves. The first two aortic arches are lost, and with the advent of pulmonary circulation, the new pulmonary arteries are derived from the sixth pair of aortic arches. Blood returning from the lungs via the pulmonary veins enters the left atrium. The branchial vessels have diminished; however, because *Necturus* is neotenic it still has the larval style of branchial respiration, with three external gills. The venous system exhibits two major changes. First, all of the systemic venous drainage comes into the right atrium via the new **postcava.** Second, the lateral abdominals are represented by the single **ventral abdominal.** The renal portal system persists, but in amniotes becomes reduced, and in mammals is completely lacking; its initial function, and its absence in mammals, are enigmatic.

Mammals

The mammalian circulatory system is adapted for high arterial pressure and efficient separation of oxygenated from unoxygenated blood. The same statement characterizes the avian circulatory system, and the similarity of the two represents an excellent example of convergent evolution associated with the acquisition of an endothermic mode of existence. In mammals the heart is completely separated into right and left sides by complete atrial and interventricular septa (see Fig. 9-1). Further reduction has occurred in the aortic arches; only III, IV, and VI are retained (see Fig. 9-2). The old anterior and common cardinals have been transformed into an anterior vena cava (precava); the postcava continues to serve posteriorly and takes on greater importance. The renal portal system is completely lost, and the lateral (or ventral) abdominal system is represented only in the embryo, as the umbilical veins.

HEART AND AORTIC ARCHES

Utilizing either the cat heart or a preserved sheep or other mammalian heart, study its structure following **Figs. 9-7** and **9-8.** If you use the cat, you will have to remove the pericardial sac and the surrounding thymus gland, which lies ventrally. The posterior two-thirds of the heart is formed of the large **left** and **right ventricles,** which are completely separated from one another by the **interventricular septum.** The **atria,** which form the anterior aspect of the heart, are also completely separated internally by the **interatrial septum.** Each atrium has a small **auricle** on its side. The more ventral of the large arteries of the base of the heart is the **pulmonary artery** or **trunk,** which arises from the right ventricle and extends to the lungs. The large, dorsal arch of the **aorta** arises from the left ventricle; two **coronary arteries** arise at its base and pass into the muscular walls of the heart. The **posterior vena cava** emerges from the diaphragm and enters the right atrium along with the **anterior vena cava** from the head region. The latter receives the coronary veins from the heart. **Pulmonary veins** from the lungs enter the left atrium.

The base of the pulmonary trunk and aortic arch is formed in part from the conus arteriosus of the fishes. The sinus venosus is incorporated into the wall of the right atrium as the **sinoatrial node;** it functions as the pacemaker.

To reveal the internal structure of the heart, it is necessary to make appropriate incisions through the atria and the ventricles. Determine the relationships of the great vessels to the different regions of the heart, including their entrances and exits from the various chambers. The atria are separated by the interatrial septum; the oval depression in the right wall of the septum is the **fossa ovalis.** This depression is what remains of the opening, the **foramen ovale,** that exists in the atrial septum during ontogeny. During the development of the fetus, blood passes through the foramen ovale di-

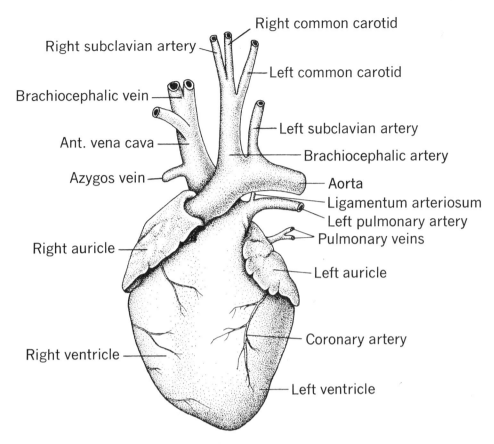

Fig. 9-7. Ventral view of the heart and associated vessels of the cat.

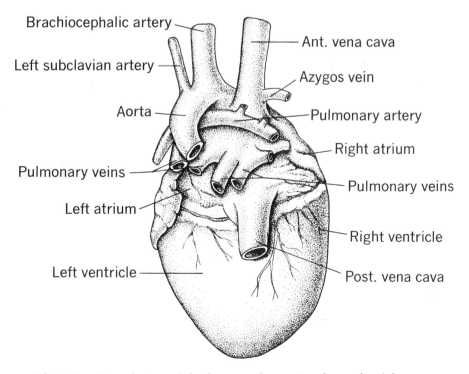

Fig. 9-8. Dorsal view of the heart and associated vessels of the cat.

rectly to the left side of the heart, since oxygenation cannot occur within the lungs; the foramen closes off at birth.

The valve guarding the opening of the ventricle is the **tricuspid** (right atrioventricular) **valve;** it has three flaps. The left atrioventricular valve consists of two flaps, and is called the **bicuspid** or **mitral valve.** On each side, the flaps of the valves are connected with the ventricular walls by tendinous cords known as the **chordae tendineae.** They also connect to mus-

cular extensions of the ventricular walls, the **papillary muscles.** These muscles prevent eversion of the valves into the atrial regions. Other valves are found in the base of the pulmonary artery (**pulmonary valve**) and the base of the aorta (**aortic valve**). These valves consist of three semilunar valves in the base of each vessel, just where it leaves the heart.

The pulmonary trunk and the arch of the aorta are the aortic arches leaving the base of the heart. The two trunks are attached after

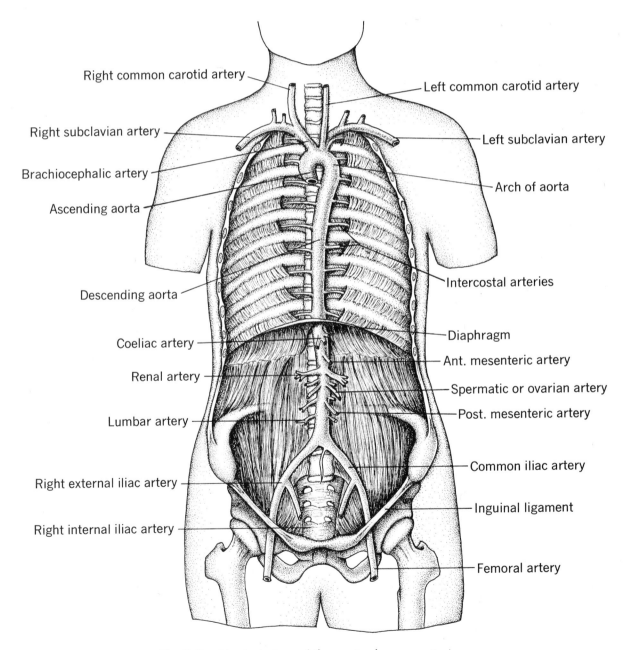

Right common carotid artery

Right subclavian artery

Brachiocephalic artery

Ascending aorta

Descending aorta

Coeliac artery

Renal artery

Lumbar artery

Right external iliac artery

Right internal iliac artery

Left common carotid artery

Left subclavian artery

Arch of aorta

Intercostal arteries

Diaphragm

Ant. mesenteric artery

Spermatic or ovarian artery

Post. mesenteric artery

Common iliac artery

Inguinal ligament

Femoral artery

Fig. 9-9. Ventral view of the major human arteries.

emerging from the heart by a ligament, the **ligamentum arteriosum,** which is the vestige of an embryonic vessel, the **ductus arteriosus.** The ductus arteriosus serves during embryogenesis as a bypass from the pulmonary arch to the arch of the aorta so that blood from the right ventricle does not go to the lungs. It closes off at birth. Slightly beyond the ligamentum arteriosum, the pulmonary trunk bifurcates into right and left pulmonary arteries. The major arteries of the head and neck emerging from the arch of the aorta are listed below (see **Figs. 9-7, 9-8,** and **9-9**).

Brachiocephalic (innominate): This is the large trunk nearest the heart; it divides shortly after emerging.

Right subclavian: This lateral branch of the brachiocephalic goes into the right pectoral appendage.

Right common carotid: This is a medial branch of the brachiocephalic that goes to the right side of the neck and head regions.

Left common carotid: This vessel passes to the left side of the neck and head regions. Both common carotids extend forward parallel to the trachea, and just before the level of the larynx divide into **internal** (more dorsal) and **external** (more ventral) **carotids.**

Left subclavian: The left subclavian extends into the left pectoral appendage.

EMBRYOLOGICAL ORIGINS OF ARTERIES

There are dozens of smaller arterial branches that extend from these major arteries; however, this list indicates the basic mammalian pattern.

During embryogenesis all six primitive pairs of the aortic arches appear; they connect the ventral aorta and the anteriorly paired dorsal aorta, just as they do in adult fishes. The principal disappearances during development include aortic arches I, II, and V, and some minor vessels in the vicinity of these. The ventral sections of arch VI give rise to the pulmonary arteries in the adult; the dorsal portion of the left segment of arch VI gives rise to the ductus arteriosus, which persists in the adult as the ligamentum arteriosum. The arch

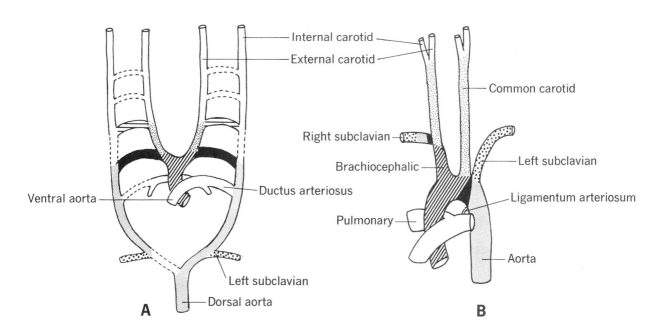

Fig. 9-10. Diagrammatic views of the ventral aspect of the aortic arches and their derivatives in man. *A,* human embryo; *B,* human adult. Compare these diagrams with *A* and *D* in Fig. 9-2. (After Barry.)

of the aorta is formed by the left segment of arch IV and the left region of the dorsal aorta. In birds the equivalent right segments of arch IV and the dorsal aorta give rise to the aortic arch; in the ectothermic groups, both systemic arches are present. The right segment of arch IV and a portion of the dorsal aorta give rise to the right subclavian. The ventral aorta forms the common carotids with the segments between arches III and IV. The dorsal aorta (paired) anterior to arch III forms the internal carotids, and the equivalent region of the ventral aorta forms the external carotids. The derivations of the mammalian aortic arches are illustrated in **Fig. 9-10.**

BRANCHES OF THE DESCENDING AORTA

Upon beginning its caudal descent into the posterior region of the body, the aortic arch is known as the descending aorta; it is the homologue of the dorsal aorta of other vertebrates, and its branches are equivalent to those described for the shark, including derivatives such as paired intersegmentals, paired lateral visceral arteries, and unpaired ventral visceral arteries. Major branches, from anterior to posterior, are listed below (and see Fig. 9-9):

Intercostals: The intercostals are a series of modified intersegmentals in the thoracic region; they send branches to the intercostal spaces, the esophagus, and the bronchi.

Phrenics: The phrenics are paired arteries to the diaphragm; they may be the last pair of intercostals.

Coeliac: This ventral visceral artery is much the same as in the shark. It sends three main branches (**splenic [lienalic]**, **gastric**, and **hepatic**) to the spleen, stomach, liver, pancreas, and part of the duodenum.

Anterior mesenteric: The anterior mesenteric is another median visceral vessel, arising posterior to the coeliac. It passes to the greater part of the intestinal region.

Renals: A pair of renals goes to the metanephroi.

Spermatics and **ovarians:** These are paired vessels that pass to the testes or ovaries.

Lumbars: A median series of lumbar arteries arises from the dorsal aorta. These vessels bifurcate and pass into the lumbar musculature. Equivalent veins arise in company with the lumbar arteries.

Posterior mesenteric: This median vessel passes both anterior and posterior along the large intestine, paralleling the posterior mesenteric vein.

External iliacs: The paired external iliacs pass laterally into the thigh. Upon leaving the abdominal cavity they are known as the **femoral arteries;** these arteries pass into the pelvic appendages to supply the deep muscles and associated structures.

Internal iliacs: These paired vessels arise from the dorsal aorta just posterior to the external iliacs. They send branches to the bladder, pelvic muscles, rectum, and reproductive organs.

Caudal: This is the extension of the aorta into the tail region.

ANTERIOR SYSTEMIC VEINS

The anterior systemic veins include those that drain into the **anterior vena cava (precava).** In the cat, only the right anterior vena cava is present. The major tributaries of the anterior vena cava are listed below (and see **Fig. 9-11**).

Azygos: The azygos veins enter the anterior vena cava on its dorsal aspect very near the heart. They receive numerous intercostal veins from the intercostal spaces, and also veins from the esophagus and bronchi. The azygos veins are homologous to the posterior cardinal veins of the lower vertebrates.

Brachiocephalics (innominates): This is the name given to the two large branches whose confluence forms the precava. Each brachiocephalic is formed by the confluence of the following:

External jugulars: The right and left external jugulars drain the head and face on either side. Small **internal jugulars** join the external jugulars somewhere near their bases, but the point of entry may vary considerably between individuals.

Subclavians: Large subclavians on either side receive blood from the shoulder regions and arms. Within the arm they are known as the **brachials.**

Many smaller tributaries contribute to the complex of the anterior systemic veins, but the above illustrate the basic mammalian pattern.

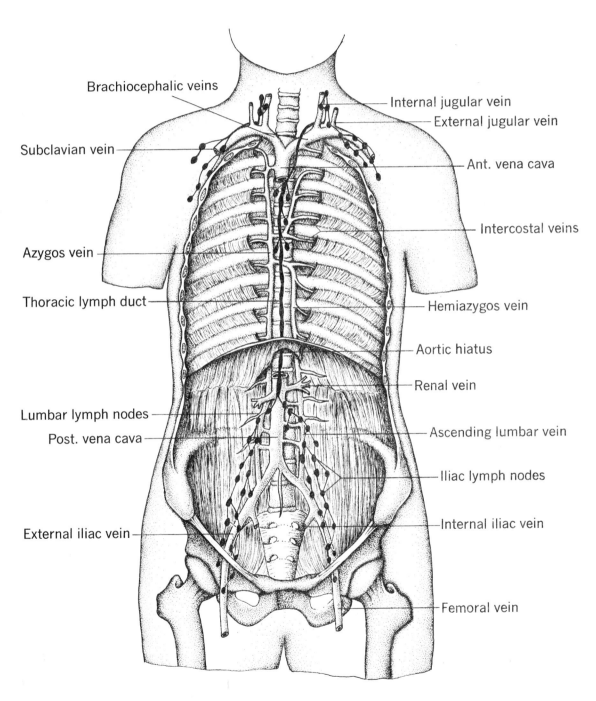

Fig. 9-11. Ventral view of the major human veins and lymphatic vessels.

POSTERIOR SYSTEMIC VEINS

The posterior systemic veins include those that drain into the **posterior vena cava (postcava)**. The tributaries of the posterior vena cava, from anterior to posterior, are listed below (and see Fig. 9-11).

Phrenics: The small, paired, phrenic veins enter the postcava as it passes through the diaphragm.

Hepatics: Several hepatics collect blood from the liver and empty into the postcava at the level of its entry into the diaphragm.

Renals, spermatics, ovarians, and **lumbars:** These veins parallel the arteries of the descending aorta.

Common iliacs: The confluence of the common iliacs forms the postcava.

External iliacs: The external iliacs receive the **femoral veins** from the thigh.

Internal iliacs: The smaller medial internal iliacs drain the pelvic muscles and pelvic visceral organs, including the urinary bladder.

Caudal: This vein enters the fork of the common iliacs.

THE HEPATIC PORTAL SYSTEM

All the veins that drain the abdominal viscera converge in the **hepatic portal vein,** which conveys this blood to the liver. This complex is termed the **hepatic portal system.** Triple-injected specimens will exhibit this system in yellow latex. The hepatic portal vein (portal vein) may be located in the lesser omentum, and traced anteriorly to the liver and posteriorly to the many vessels contributing to it.

10

The Urogenital System

The urogenital system consists of the kidneys, the gonads, and the ducts of both systems. There is no real functional link between the kidneys and gonads; however, due to their similar developmental origins and their frequent sharing of common ducts, they are often treated together as one system. Kidneys function in water balance and in the removal of nitrogenous waste products, while the gonads function only in a reproductive capacity.

The kidneys and kidney tubules arise from the nephrotomal plate or mesomere (see Fig. 6-1), a longitudinal ribbon of mesoderm that in the embryo extends from the anterior abdominal region to the cloaca. Kidney tubules develop anteriorly as one pair per body somite, but posteriorly, tubules become more abundant. Longitudinal ducts develop first as extensions of the anterior kidney tubules; these ducts extend posteriorly until they meet the cloaca. In the primitive condition, the kidney is thought to have extended the entire length of the intermediate mesoderm, with segmentally arranged tubules drained by a longitudinal duct extending from the cranial end of the kidney to the cloaca. Such a hypothetical kidney is termed an archinephros or holonephros (**Fig. 10-1**), and its longitudinal duct an archi- or holonephric duct. The ducts of kidneys are named according to the type of kidney they drain. Though largely hypothetical, the archinephric kidney is approximated in some of the larval cyclostomes.

Early in the ontogeny of amniotes, the intermediate mesoderm extends the entire length of the coelom. At first, segmentally arranged kidney tubules develop in the anterior region only. This pronephros degenerates and is succeeded by the mesonephros (developed in the middle region of the nephrotomal plate), which later degenerates and gives way to the metanephros of the amniotes, a kidney developed by a budding-off of the posteriormost nephrogenic tissue. The metanephros has millions of non-segmentally arranged kidney tubules, and is drained by a new duct called the ureter. In amniotes, the adult kidney is termed an opisthonephros, and its duct is most appropriately termed an opisthonephric duct; this kidney occupies all of the nephrotomal plate except the most anterior portion; opisthonephroi are preceded in ontogeny by pronephroi.

It should be pointed out here that the naming of the kidney types is somewhat arbitrary; there are no black-and-white distinctions between the different types. Much confusion has arisen because of the terminology used. For example, many authors use the term mesonephros for the anamniote type of kidney, and term its duct the Wolffian duct. Other authors use the term opisthonephros for the anamniote type, and term its duct the archinephric duct; such nomenclature has led to great confusion. However, the important thing to remember is that all the ducts, from the archinephric

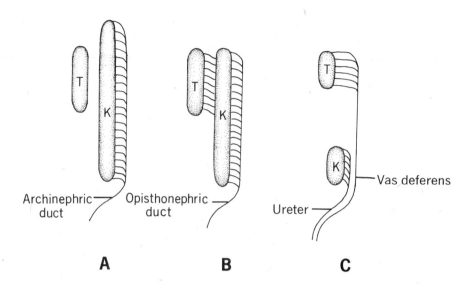

Fig. 10-1. Diagrammatic illustration showing the evolution of the urogenital ducts in male vertebrates (testes, *T*; kidney, *K*). In the hypothetical ancestral holonephros (*A*), the kidney forms along the entire nephrotomal plate, and the archinephric duct has a purely urinary function; gametes erupt into the coelom and exit via abdominal pores. In most anamniotes (*B*), the opisthonephros (derived from all but the anteriormost portion of the nephrotomal plate) is drained by the opisthonephric (= archinephric) duct, but efferent ductules from the testes carry sperm to the kidney duct, which transports both sperm and urine. In amniotes (*C*), the metanephric kidney arises as a bud displaced laterally from the posterior portion of the nephrotomal plate; the kidney is drained by a new duct, the ureter, and the old archinephric duct (= Wolffian duct, ductus or vas deferens) is used exclusively for the transport of sperm.

through the meso- and opisthonephric types, are homologous; the ureter is an innovation and develops from an outgrowth of the posterior region of the embryonic duct.

In the primitive condition, the gametes of both sexes are discharged into the coelom and to the exterior via abdominal pores (see Fig. 10-1); this is the condition in the lamprey. Later, ducts develop to transport the gametes. By the early anamniote stage, the kidney duct takes over the function in the male. Modified anterior kidney tubules, called efferent ductules (vasa efferentia), transport the sperm from the testes to the kidney ducts. In mammals the kidney duct divorces itself from kidney function and becomes solely concerned with sperm transport; it is the ductus deferens (vas deferens), modified around the testes as the epididymis. In the embryo of the female, two longitudinal ducts, the Muellerian ducts, develop ventrally, but in close proximity to the

developing kidneys. These ducts degenerate for the most part in the male, but in the female give rise to the entire tract: ostium, oviduct, uterus, and vagina (see Fig. 10-9).

The gonads (testes and ovaries) develop from paired genital ridges medial to the kidneys (see Fig. 6-1). Gonads are undifferentiated at an early embryonic stage and give rise to either ovaries or testes; in each case a different region predominates in development.

Fishes

The urogenital system of the elasmobranchs is an ideal system for providing insight into the basic anamniote condition; it is, however, specialized in some respects. In both sexes the kidneys are opisthonephroi and are drained by opisthonephric ducts (with an accessory opis-

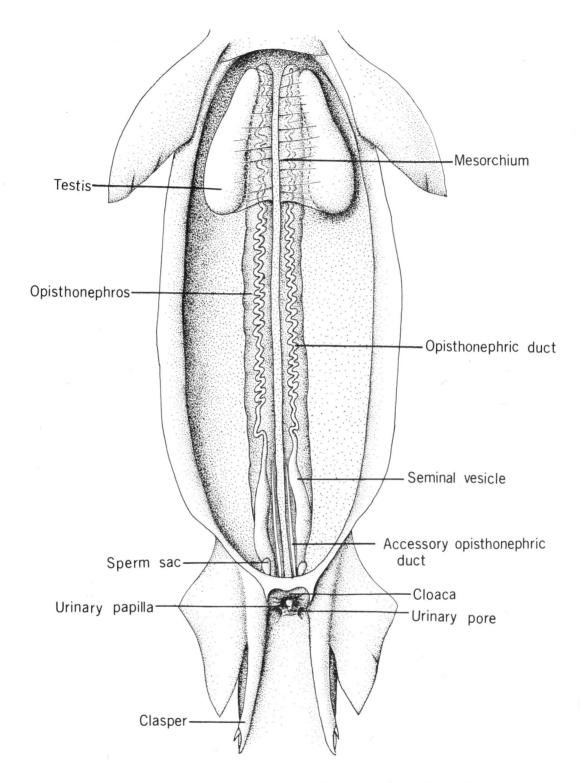

Fig. 10-2. Ventral view of the urogenital system of the male *Squalus*.

thonephric duct in the male). Determine the sex of your specimen (recall that the male has claspers), and study the urogenital systems following **Figs. 10-2** and **10-3.**

The **opisthonephroi** are dark flat bodies lying in the dorsal body wall immediately lateral to the dorsal aorta, and extending from the ante-

rior region of the liver to the level of the cloaca. They are retroperitoneal, lying dorsal to the parietal peritoneum. In the male the **opisthonephric ducts** appear as a pair of coiled ducts on the ventral surface of the kidney. Urine is transported from the kidney to the ducts by microscopic opisthonephric tubules. One may

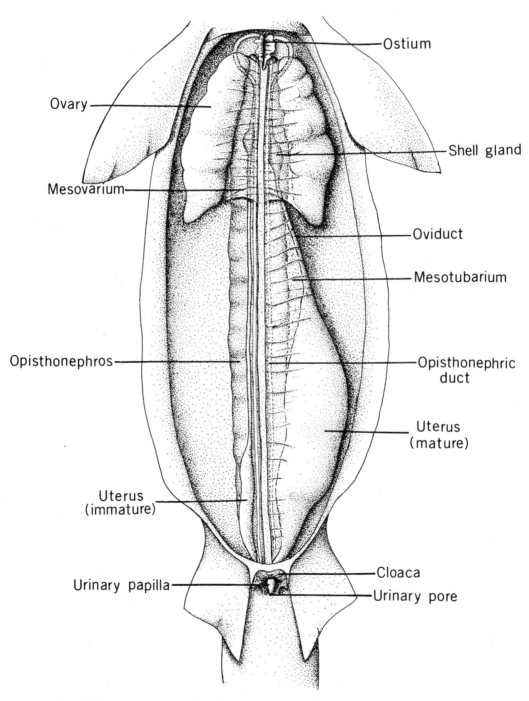

Fig. 10-3. Ventral view of the urogenital system of the female *Squalus*.

cut through the parietal peritoneum to trace the opisthonephric ducts. The convoluted ducts become expanded posteriorly as the **seminal vesicles,** which have small posterior evaginations, the **sperm sacs.** The **cloaca** may be split open with a scalpel to determine the various openings. The two seminal vesicles unite posteriorly and open into the **urogenital sinus,** within the **urogenital papilla.** In most anamniotes the ducts of the kidneys carry both urine and sperm. In the dorsal body wall two **testes** are suspended into the coelom by a mesentery, the **mesorchium.** Small modified tubules, the **efferent ductules,** carry sperm from the testes to the opisthonephric duct. Fertilization is internal

in the cartilaginous fishes; the sperm passes from the cloaca to a groove on the dorsal aspect of the **clasper;** the clasper is inserted into the cloaca of the female.

In the female the opisthonephric ducts are smaller than those of the male, are uncoiled, and are partially embedded in the tissue of the opisthonephroi. The opisthonephric ducts empty into the urinary sinus, within the papilla. The paired **ovaries** are suspended in the dorsal body wall by mesenteries, the **mesovaria.** Upon ovulation, the ova break away from the ovary and are shunted directly into the abdominal cavity; from there they enter the ostium.

In both sexes a pair of **Muellerian ducts** ap-

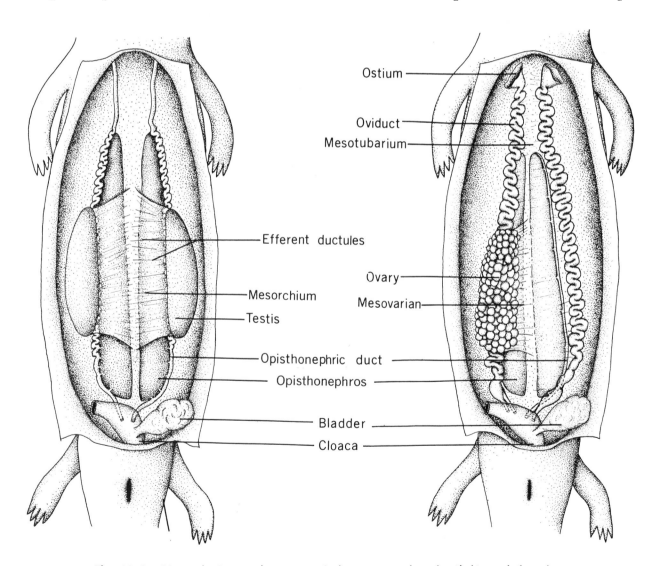

Fig. 10-4. Ventral views: the urogenital systems of male (*left*) and female (*right*) Necturus.

pear ventral to the kidneys early during embryogenesis. In the male they fail to develop fully, and give rise to very minor structures. However, in the female they give rise to the reproductive tracts. Trace the tracts from anterior to posterior. The **ostium** shunts the eggs into the oviducts. An expansion of the oviduct, the **shell gland,** secretes a temporary shell around the egg. The **oviducts** lead into the **uteri,** which are suspended in the body cavity by mesenteries, the **mesotubaria.** *Squalus* is ovoviviparous, and the "pups" develop within the uterus.

Amphibians

Obtain a specimen of **Necturus** for dissection of the kidneys and gonads (**Fig. 10-4**).

Amphibians still possess the basic anamniote type of kidney, the **opisthonephros.** In both sexes the kidneys appear as elongated, brown organs located primarily in the posterior region of the abdominal cavity; in the male they are slightly more developed anteriorly, since they serve a genital function in the transport of sperm. In the female the **opisthonephric ducts** are small, uncoiled tubes running along the lateral borders of the nephroi; they empty separately into the cloaca. In males the ducts are conspicuous coiled structures on the lateral aspect of the kidney; they may extend anteriorly to the cranial border of the kidneys. As in the female, the ducts empty separately into the **cloaca.** The **testes** are located near the anterior ends of the kidneys; they are suspended in the abdominal cavity by the **mesorchium. Efferent ductules** carry sperm from the testes through the mesorchium to the modified kidney tubules. There are several glands in the male cloaca, whose secretions serve to agglutinate sperm into small packages, called **spermatophores.** These are placed in the water for the female to

pick up, or in some cases are deposited into the cloaca of the female.

In the female the conspicuous **ovaries** contain eggs in various degrees of development. Equally conspicuous are the large convoluted **oviducts,** which extend nearly the entire length of the abdominal cavity. Upon ovulation, the eggs break away from the ovaries and are engulfed by the **ostia;** from there they travel down the oviducts to the cloaca. There is a slight expansion of the oviducts posteriorly. This is sometimes called a **uterus;** however, recall that in amphibians fertilization is external, and the young develop outside the female reproductive tract.

In both the male and female specimens there is a conspicuous **urinary bladder** that develops as an outpocketing of the cloaca; this is a feature first appearing in the amphibia.

Mammals

Obtain a specimen of a cat for dissection of the mammalian kidneys and gonads (see **Figs. 10-5 through 10-9**).

A new type of kidney evolved with the emergence of the amniotes. The amniote kidney is derived from only the posterior portion of the embryonic nephrotomal plate; a small metanephric bud migrates laterally to give rise to the new **metanephros.** In mammals the old opisthonephric (mesonephric) duct is taken over entirely by the male genital structures, and the kidneys in both male and female are drained by a new structure, the **ureter.** The metanephroi are retroperitoneal structures, lying outside the parietal peritoneum. They exhibit a medial concavity, the **hilum.** Upon cleaning the area of the hilum, one can observe the ureter and the renal arteries and veins; these are easily traced back to the dorsal aorta. In frontal section one may observe that the kidney is clearly divided into a light, outer **cortex,** and a darker, inner **medulla.** Embedded in fat anteromedial to each kidney is an **adrenal gland** (Figs. 10-5 and 10-6). In primitive tetrapods the adrenal glands are usually in the form of clusters of cortical (outer) and medullary (inner) cells in patches along the ventral surface of the kidneys. In mammals they are discrete bodies with the medullary cells located centrally.

After studying the anatomy of the kidney, you should trace one of the ureters posteriorly beneath the peritoneum. Just before entering the urinary bladder, the ureters pass dorsal to the uterine horns in the female and the vas deferens in the male. By cutting open the bladder, one can observe the points of entry of the ureters, and the point of exit of the urethra, which carries urine to the exterior.

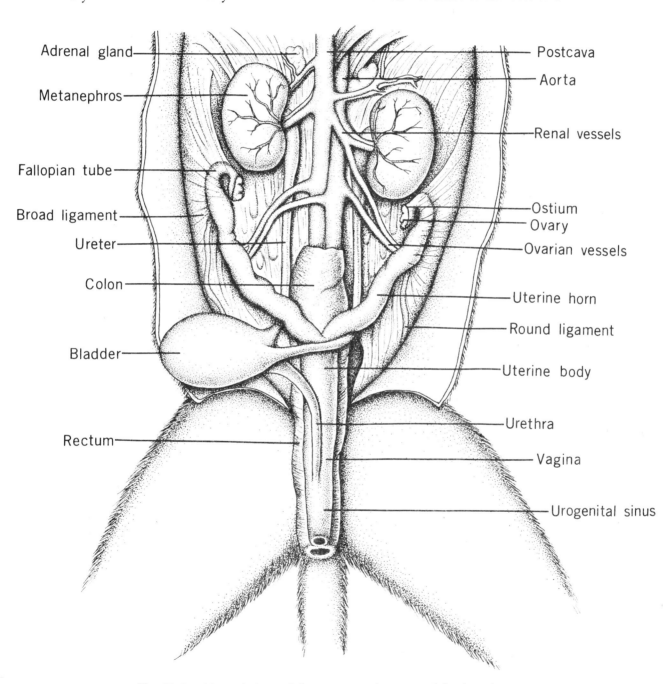

Fig. 10-5. Ventral view of the urogenital system of the female cat.

FEMALE

The paired **ovaries** of the female lie just caudal to the kidneys; they are small structures approximately 1 cm long. **Mesovaria** suspend the ovaries into the body cavity. Anteriorly, each ovary is surrounded by a pocket of mesentery housing the anterior portion of the oviduct called the **fallopian tube.** This tube continues from its tiny opening into the coelom called

the **ostium.** Only with care can the ostium be observed. Fertilization occurs in the fallopian tubes; the egg then passes down into the **uterine horn,** where implantation in the uterine lining occurs. The uterine horns (along with the fallopian tubes) are suspended by a mesentery, the **broad ligament.** The uterine horns merge caudally into a single uterine body. By making a longitudinal incision into the body of the uterus, one can determine the relation-

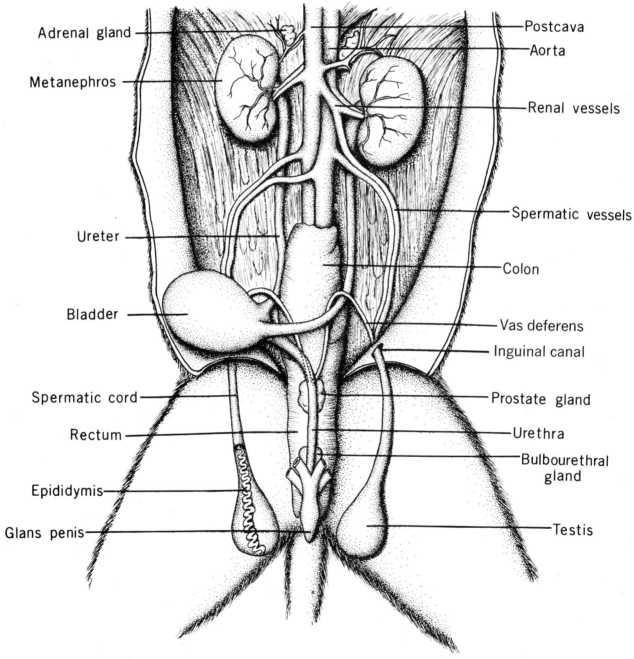

Fig. 10-6.   Ventral view of the urogenital system of the male cat.

ship of the uterus to the caudad **vagina**: the posterior end of the uterus extends slightly into the vagina as the **cervix**; it contains a small opening.

In order to observe the caudal portion of the genital tract in the pelvic cavity, it is necessary to remove the muscles in the area of the pelvis and cut through the pubic symphysis, thus exposing the caudal end of the tract. The **urethra** (coming from the bladder) joins the caudal end of the genital tract to form the **urogenital sinus**, which leads to the exterior. By slitting the wall of the urogenital sinus, one can observe the fine structures. In the ventral wall of the urogenital sinus is the small, rodlike **clitoris**; it is formed of erectile tissue and is the sexual homologue of the penis in the male. The folds that bound the opening of the urogenital sinus are collectively called the **vulva**; included in this are the lateral lips of the opening, the **labia majora.**

MALE

By carefully incising the **scrotal sac** on one side, one may expose the **testis**, the sexual homologue of the female ovary. Testes serve not only in the production of spermatozoa, but also in important endocrinological functions through the production of testosterone. By making an incision just beneath the skin lateral to the penis (as a continuation of the previous incision), one may encounter the **spermatic cord,** which contains all of the structures entering the testis. The cord consists of its external covering, the **tunica vaginalis,** the **internal spermatic artery** and **spermatic vein,** the **spermatic nerves,** the **vas deferens,** and the **cremaster muscle,** which is derived from part of the internal oblique. This muscle serves to retract the testes, slightly in some mammals, or completely into the abdominal cavity in others. Note that as the cord courses craniad it passes through the abdominal musculature; the opening is called the **external inguinal ring.** The ring leads into the **inguinal canal,** and opens into the peritoneal cavity by the **internal inguinal ring.** You should now be able to understand what occurs in an inguinal hernia.

The testes were originally, in the embryo and immature animal, within the abdominal cavity; upon maturation they migrated into the scrotal sac, taking with them their surrounding peritoneum. Note that the actual scrotum is separated by a median septum and consists of, in addition to the outer layer of skin,

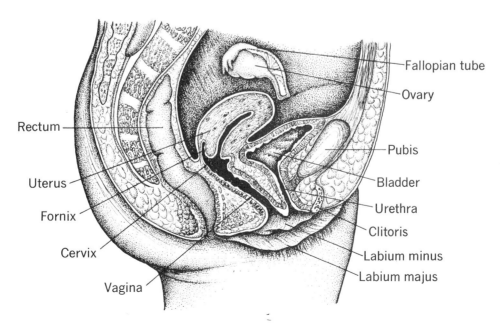

Fig. 10-7.   Sagittal section of the human female urogenital system.

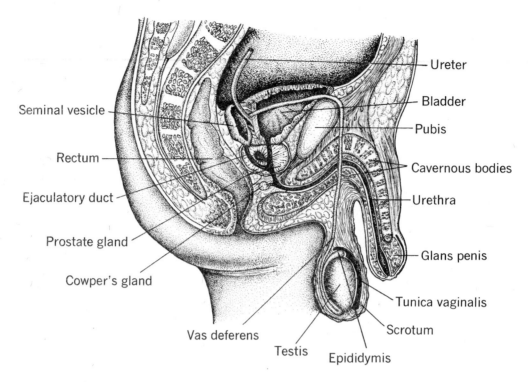

Fig. 10-8. Sagittal section of the human male urogenital system.

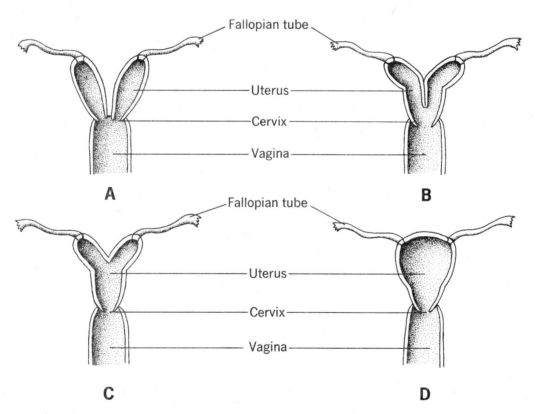

Fig. 10-9. Diagrammatic illustrations of the uteri of various placental mammals to show the degree of fusion of the embryonic Muellerian ducts. *A*, duplex uterus (rodents and rabbits); *B*, bipartite uterus (carnivores); *C*, bicornuate uterus (ungulates); and *D*, simplex uterus (primates).

a deeper pair of cordlike sacs, which, along with the testes, have entered the skin sacs. As the testes descend into the scrotal sac they must carry with them two layers of peritoneum, their own visceral layer (of the tunica vaginalis), and that which lines the sac itself, the parietal layer (of the tunica vaginalis). Thus, the spermatic cord is actually a blind-ended pouch, the end of which contains the testes. The **mesorchium** suspends the testes and part of the spermatic cord. The posterior area of the mesorchium is thickened as a band of tissue called the **gubernaculum.** Before parturition, the gubernaculum shortens, causing (at least in part) the descent of the testes. At the caudal region of the testes the **vas deferens** becomes transformed into the **epididymis,** a coiled tube connected to the testes by efferent ductules. The ductules are homologous to the modified kidney tubules of the anterior end of the opisthonephric kidney. In addition, the vas deferens is the homologue of the old opisthonephric duct (also called the Wolffian duct). Recall that this duct was initially a duct for urine only, later was utilized for the transport of both urine and sperm, and in mammals serves only for the transport of sperm. The coiled epididymis thus serves as a pathway for sperm as they leave the testes.

In order to study the remainder of the male urogenital system it will first be necessary to split the pubic symphysis and remove sufficient bone and tissue for complete exposure. One should now trace the **vasa deferentia** to the point where they independently enter the **urethra.** A small **prostate gland** is present at this point; it is one of the accessory sex glands that contributes to the seminal fluid. A pair of **bulbourethral (Cowper's) glands** enter the urethra by small ducts dorsal to the root of the penis. The **penis** (intromittent organ) is covered by skin that also sheaths its enlarged distal end, the **glans penis;** the sheath is called the **prepuce.** Make a cross section of the body of the penis and observe the two bilateral bodies that comprise most of it. These are called the **corpora cavernosa,** and are composed of erectile tissue with labyrinthine vascular channels that become engorged with blood upon erection. A midventral body of spongy tissue, the **corpus spongiosum,** surrounds the urethral canal.

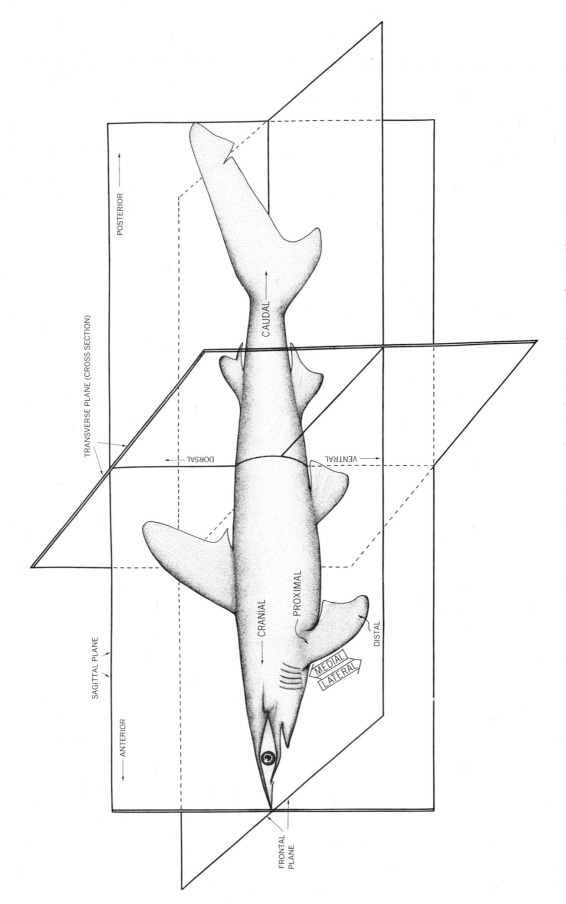

Fig. A-1. A diagram of the bonnet-headed shark (*Sphyrna tiburo*) illustrating anatomical directions and planes.

APPENDIX 1

Terminology for Vertebrate Dissection

The terms used in anatomical dissection vary somewhat from one book to another. The following are commonly used terms for anatomical directions, planes, sections, and axes.

IMPORTANT ADJECTIVES USED IN ANATOMY

(1) **Anterior** (opposite = **posterior**): pertaining to the head end of an animal. **Posterior:** pertaining to the tail end. These two terms may apply to an organism or to a bone, organ, or organ system.

(2) **Dorsal** (opposite = **ventral**); pertaining to the upper surface, back, or spinal side of an organism or organ. **Ventral:** pertaining to the under side or belly of an organism or organ.

(3) **Proximal** (opposite = **distal**): close to, near, or proximate, referring to a point of attachment to the body. **Distal:** referring to a point toward the free end of the structure under consideration.

(4) **Median, medial,** or **mesial** (opposite = **lateral**): lying in the plane that divides the body into two bilateral halves; lying in the "middle" plane. **Lateral:** lying toward the sides of the organisms, or away from the median plane in either direction. (Note that the words *median, medial,* and *mesial* are used in a similar manner.)

(5) **Cranial** (opposite = **caudal**): pertaining to the skull, or skull area. **Caudal:** pertaining to the tail region. (Note that the word *cephalic* refers to the head region, but not specifically to the skull.)

(6) **Superficial** (opposite = **deep**): near the surface. **Deep:** away from the surface.

(7) **Superior** (opposite = **inferior**): of erect animals (e.g., man), used instead of *anterior*. **Inferior:** of erect animals, used instead of *posterior*.

IMPORTANT NOUNS USED IN ANATOMY

(1) **Cranium** = skull
(2) **Cephalon** = head
(3) **Cauda** (pl., **caudae**) = tail
(4) **Dorsum** = upper surface
(5) **Venter** = lower surface

IMPORTANT ADVERBS USED IN ANATOMY

(1) **Cephalad** = toward the head.
(2) **Craniad** = toward the skull.
(3) **Caudad** = toward the tail.
(4) **Distad** = toward a distal end.
(5) **Proximad** = toward a proximal end.
(6) **Dorsad** = toward the dorsum or upper surface.
(7) **Ventrad** = toward the venter or lower surface.

(8) **Laterad** = toward the sides.

(9) **Mediad** = toward the middle.

ANATOMICAL AXES

(1) **Anterior-posterior axis** (= **Longitudinal axis**).

(2) **Dorso-ventral axis** (= **sagittal axis**).

(3) **Right-left axis** (= **mediolateral** or **transverse axis**).

ANATOMICAL PLANES AND SECTIONS

Planes are surfaces that are constructed by two axes.

(1) **Transverse plane**—constructed by dorso-ventral and lateral axes. A cut through a transverse plane results in a **transverse section,** commonly referred to as a **cross section.**

(2) **Frontal plane**—constructed by longitudinal (antero-posterior) and lateral axes. A cut through a frontal plane results in a **frontal section.**

(3) **Sagittal plane**—constructed by longitudinal and dorso-ventral axes. A cut through a sagittal plane results in a **sagittal section.**

(4) **Parasagittal plane**—constructed as a sagittal plane, but lateral to the true sagittal plane (to the left or right side). A cut through a parasagittal plane results in a **parasagittal section.**

APPENDIX 2

Classification of the Vertebrates

The following is an abbreviated classification of the major groups of vertebrates. It includes all of the groups discussed in the text.

PHYLUM CHORDATA

Subphylum Hemichordata[1]—
acorn worms and pterobranchs

Subphylum Urochordata—
tunicates

Subphylum Cephalochordata—
lancelets

Subphylum Vertebrata—
vertebrates

CLASS AGNATHA—jawless fishes

    ORDERS HETEROSTRACI, OSTEOSTRACI, ANASPIDA, COELOLEPIDA—"ostracoderms"

    ORDER CYCLOSTOMATA—living jawless fishes, the cyclostomes

        SUBORDER PETROMYZONTIDA—lampreys

        SUBORDER MYXINOIDEA—hagfishes

[1] The beardworms (Pogonophora) are often treated as a completely separate group, and the hemichordates as a separate phylum.

* CLASS PLACODERMI—archaic armored jawed fishes

CLASS CHONDRICHTHYES—cartilaginous fishes

    SUBCLASS ELASMOBRANCHII—sharks, skates, and rays

        ORDER SELACHII—typical sharks

        ORDER BATOIDEA—skates and rays

    SUBCLASS HOLOCEPHALI—chimaeras

        ORDER CHIMAERIFORMES—chimaeras or rat fishes

CLASS OSTEICHTHYES—higher bony fishes

    SUBCLASS ACTINOPTERYGII—ray-finned fishes

        ORDER CHONDROSTEI—primitive ray-finned fishes; extinct "paleoniscoids" (living paddlefish, *Polyodon*, and sturgeon, and the Nile bichir, *Polypterus*)

        ORDER HOLOSTEI—primitive ray-finned fishes arising in the Mesozoic Era (the bowfin (*Amia*) and gar pikes)

        ORDER TELEOSTEI—modern bony fishes

    SUBCLASS SARCOPTERYGII (CHOANICHTHYES)—lobe-finned fishes

        ORDER CROSSOPTERYGII—coelacanths (all extinct except *Latimeria*), and the rhipidistians, which were ances-

* Extinct.

tral to the amphibia, but are now completely extinct

ORDER DIPNOI—lungfishes (extinct, except for three living genera living in South America, Africa, and Australia)

CLASS AMPHIBIA—the first tetrapods

* SUBCLASS LABYRINTHODONTIA—the first amphibians, mainly armored

SUBCLASS LISSAMPHIBIA—modern amphibians

ORDER ANURA—frogs and toads

ORDER URODELA—salamanders and newts

ORDER APODA (Gymnophiona)—caecilians (scaled, burrowing, wormlike amphibians)

CLASS REPTILIA—the first truly terrestrial tetrapods

SUBCLASS ANAPSIDA—reptiles with no temporal opening in the skull

ORDER COTYLOSAURIA—"stem reptiles"

ORDER CHELONIA—turtles

SUBCLASS LEPIDOSAURIA—reptiles with two temporal openings in the skull (includes all the diapsids except the archosaurs)

* ORDER EOSUCHIA—ancestral diapsids

ORDER RHYNCHOCEPHALIA—includes the living tuatara (Sphenodon) of New Zealand

ORDER SQUAMATA—snakes and lizards

SUBCLASS ARCHOSAURIA—ruling reptiles of the Mesozoic Era (all diapsids)

* ORDER THECODONTIA—ancestral archosaurs

ORDER CROCODILIA—crocodiles and alligators

* ORDER SAURISCHIA—large carnivorous and large amphibious dinosaurs, with triradiate pelvises

* ORDER ORNITHISCHIA—herbivorous dinosaurs, including the duckbills,

armored and horned forms, with tetraradiate birdlike pelvises

* ORDER PTEROSAURIA—flying reptiles

* SUBCLASS EURYAPSIDA—extinct groups of Mesozoic reptiles with one temporal opening in the skull; includes many marine reptiles such as the plesiosaurs and ichthyosaurs

* SUBCLASS SYNAPSIDA—reptiles with one lateral temporal opening in the skull (differ from euryapsids in position of opening)

* ORDER PELYCOSAURIA—fin-backed reptiles

* ORDER THERAPSIDA—mammal-like reptiles closely allied to the "stem reptiles"

CLASS AVES—birds (derived from the thecodont archosaurs)

* SUBCLASS ARCHAEORNITHES—represented only by the first bird, Archaeopteryx

* SUBCLASS NEORNITHES—modern birds

* SUPERORDER ODONTOGNATHAE—Cretaceous toothed birds

SUPERORDER NEOGNATHAE—all living birds

CLASS MAMMALIA—warm-blooded tetrapods, derived from the therapsid reptiles

SUBCLASS PROTOTHERIA—egg-laying mammals

ORDER MONOTREMATA—living duckbill platypuses and spiny anteaters (Echidna) of Australia and New Guinea

SUBCLASS THERIA—viviparous mammals

INFRACLASS METATHERIA—mammals bearing live but immature young, that are nourished in a pouch

ORDER MARSUPIALIA—marsupials or pouched mammals

INFRACLASS EUTHERIA—placental mammals

References

GENERAL

Abramson, D. I., ed. 1962. Blood vessels and lymphatics. New York, Academic Press.

Allison, A. C. 1953. The morphology of the olfactory system in the vertebrates. Biol. Rev. 28: 195–244.

Arey, L. B. 1965. Developmental anatomy. 7th ed. Philadelphia, W. B. Saunders Company.

Balinsky, B. I. 1965. An introduction to embryology. 2nd ed. Philadelphia, W. B. Saunders Company.

Ballard, W. W. 1964. Comparative anatomy and embryology. New York, The Ronald Press Company.

Barclay, A. E., K. J. Franklin, and M. M. L. Prichard. 1945. The foetal circulation. Springfield, Illinois, Thomas Company.

Carter, G. S. 1967. Structure and habit in vertebrate evolution. Seattle, University of Washington Press.

Colbert, E. H. 1969. Evolution of the vertebrates. 2nd ed. New York, John Wiley & Sons.

DeBeer, G. R. 1937. The development of the vertebrate skull. New York, Oxford University Press.

DeHaan, R. L. 1965. Morphogenesis of the vertebrate heart. *In:* Organogenesis. DeHaan, R. L., and H. Ursprung, eds. New York, Holt, Rinehart, and Winston.

Edgeworth, F. H. 1935. The cranial muscles of vertebrates. London, The Macmillan Company.

Fraser, E. A. 1950. The development of the vertebrate excretory system. Biol. Rev. 25:159–187.

Goodrich, E. S. 1958. Studies on the structure and development of vertebrates. New York, Dover Publications.

Grassé, P. P., ed. 1948–1958. Traité de zoologie. Vols. 1–17. Paris, Masson et Cie.

Gregory, W. K. 1951. Evolution emerging. New York, The Macmillan Company.

Halstead, L. B. 1968. The pattern of vertebrate evolution. San Francisco, W. H. Freeman & Company.

Hildebrand, M. 1974. Analysis of vertebrate structure. New York, John Wiley & Sons.

Hörstadius, S. 1950. The neutral crest. New York, Oxford University Press.

Hughes, G. M. 1963. Comparative physiology of vertebrate respiration. Cambridge, Harvard University Press.

Huxley, H. E. 1969. The mechanism of muscle contraction. Science 164:1356–1366.

Hyman, L. H. 1942. Comparative vertebrate anatomy. 2nd ed. Chicago, University of Chicago Press.

Jaeger, E. C. 1955. A source-book of biological names and terms. 3rd ed. Springfield, Illinois, Charles C. Thomas Company.

Jollie, M. 1962. Chordate morphology. New York, Reinhold Publishing Corporation.

Kappers, C. U. A., C. G. Huber, and E. C. Crosby. 1936. The comparative anatomy of the nervous system, including man, 2 vols. New York, The Macmillan Company.

Keeton, W. T. 1972. Biological science. 2nd ed. New York, W. W. Norton & Company.

Kent, G. C. 1973. Comparative anatomy of the vertebrates. 3rd. ed. St. Louis, C. V. Mosby Company.

Kingsley, J. S. 1917. Outlines of comparative anatomy of vertebrates. 2nd ed. Philadelphia, P. Blakiston's Son & Co.

Manley, G. A. 1972. A review of some current concepts of the functional evolution of the ear in terrestrial vertebrates. Evolution 26:608–621.

Mayerson, H. S. 1963. The lymphatic system. Sci. Amer. 208(6):80–90.

Nelsen, O. E. 1953. Comparative embryology of the vertebrates. New York, The Blakiston Company.

Olson, E. 1971. Vertebrate paleozoology. New York, John Wiley & Sons.

Orr, R. T. 1971. Vertebrate biology. 3rd. ed. Philadelphia, W. B. Saunders Company.

Owen, R. 1866–1868. On the anatomy of vertebrates. 3 vols. London, Longmans, Green & Co.

Parker, T. J., and W. A. Haswell. 1962. A text-book of zoology. Vol. 2. 7th ed., revised by A. J. Marshall. London, The Macmillan Company.

Parsons, T. S. (organizer). 1968. Functional morphology of the heart of vertebrates. Amer. Zoologist 8:177–229.

Petras, J. M., and C. R. Noback, eds. 1969. Comparative and evolutionary aspects of the vertebrate central nervous system. Ann. N.Y. Acad. Sci. 167:1–513.

Polyak, S. 1957. The vertebrate visual system. Chicago, University of Chicago Press.

Rensch, B. 1960. Evolution above the species level. New York, Columbia University Press.

Romer, A. S. 1958. The vertebrate story. Chicago, University of Chicago Press.

Romer, A. S. 1966. Vertebrate paleontology. 3rd ed. Chicago, University of Chicago Press.

Romer, A. S. 1967. Major steps in vertebrate evolution. Science, 158:1629–1637.

Romer, A. S. 1970. The vertebrate body. 4th ed. Philadelphia, W. B. Saunders Company.

Romer, A. S. 1972. The vertebrate as a dual animal —somatic and visceral. In: Evolutionary biology. Vol. 6 (Dobzhansky, T., M. K. Hecht, and W. C. Steere, eds. New York, Appleton-Century-Crofts.

Rouiller, C. 1962. The liver. New York, Academic Press.

Smith, H. M. 1960. Evolution of chordate structure. New York, Holt, Rinehart and Winston.

Smith, H. W. 1951. The kidney. London, Oxford University Press.

Smith, H. W. 1953. From fish to philosopher. Boston, Little, Brown and Company.

Torrey, T. M. 1971. Morphogenesis of the vertebrates. 3rd. ed. New York, John Wiley & Sons.

Van Bergeijk, W. A. 1966. Evolution of the sense of hearing in vertebrates. Amer. Zoologist 6:371–377.

Villee, C. A., W. F. Walker, Jr., and F. E. Smith. 1968. General zoology 3rd ed. Philadelphia, W. B. Saunders Company.

Walker, W. F., Jr. 1970. Vertebrate dissection. 4th ed. Philadelphia, W. B. Saunders Company.

Walls, G. L. 1942. The vertebrate eye and its adaptive radiation. Cranbrook Inst. Sci. Bull. 19.

Waterman, A. J., ed. 1971. Chordate structure and function. New York, The Macmillan Company.

Webster, D., and M. Webster. 1974. Comparative vertebrate morphology. New York, Academic Press.

Weichert, C. K. 1967. Elements of chordate anatomy. 3rd ed. New York, McGraw-Hill Book Company.

Weiss, P. 1955. Nervous system. In: Analysis of development. (Willier, B. H., Weiss, and Hamburger, eds.) Philadelphia, W. B. Saunders Company.

Wilkins, L. 1960. The thyroid gland. Sci. Amer. 202:119–129.

Wood, J. E. 1968. The venous system. Sci. Amer. 218:86–96.

Wright, M. E. 1951. The lateral line system of sense organs. Quart. Rev. Biol. 26:264–280.

Wurtman, R. J., and J. Axelrod. 1965. The pineal gland. Sci. Amer. 213:50–60.

Yapp, W. B. 1965. Vertebrates: their structure and life. New York, Oxford University Press.

Young, J. Z. 1962. The life of vertebrates. 2nd ed. London, Oxford University Press.

PROTOCHORDATES

Barrington, E. J. W. 1965. The biology of hemichordata and protochordata. San Francisco, W. H. Freeman & Company.

Berrill, N.J. 1955. The origin of vertebrates. Oxford University Press.

Bone, Q. 1960. The origin of chordates. J. Linnean Soc. London 44:252–269.

Cavey, M. J., and R. A. Cloney. 1972. Fine structure and differentiation of ascidian muscle. I.: Differentiated caudal musculature of *Distaplia occidentalis* tadpoles. J. Morphol. 138:349–373.

DeBeer, G. R. 1940. Embryos and ancestors. Oxford, Clarendon Press.

Denison, R. H. 1971. The origin of the vertebrates: a critical evaluation of current theories. Proc. North Amer. Paleontol. Convention, Pt. H: 1132–1146.

Fell, H. B. 1948. Echinoderm embryology and the origin of chordates. Biol. Rev. 23:81–107.

Garstang, W. 1928. The morphology of the Tunicata, and its bearing on the phylogeny of the Chordata. Quart. J. Micr. Sci. 72:51–187.

Grassé, P. P., ed. 1948–1958. Traité de zoologie. Vol. 11. Paris, Masson et Cie.

Gregory, W. K. 1946. The roles of motile larvae and fixed adults in the origin of the vertebrates. Quart. Rev. Biol. 21:348–364.

Hyman, L. H. 1959. The invertebrates: smaller coelomate groups, Chaetognatha, Hemichordata, Pogonophora, Phoronida, Ectoprocta, Brachiopoda, Sipunculida, the coelomate Bilateria. Vol. 5. New York, McGraw-Hill Book Company.

Ivanov, A. 1955. Pogonophora. (Translated by A. Petrunkevitch.) Syst. Zool. 4:170–178.

Jollie, M. 1973. The origin of chordates. Acta Zool. 54:81–100. (This article challenges most of the current theories.)

Moller, P. C., and C. W. Philpott. 1973. The circulatory system of *Amphioxus* (*Branchiostoma floridae*). I.: Morphology of the major vessels of the pharyngeal region. J. Morphol. 139:389–406.

Needham, J., and D. M. Needham. 1932. Biochemical evidence regarding the origin of vertebrates. Sci. Progress 26:626.

Nichols, D. 1967. The origin of echinoderms. Zool. Soc. of London Symp. 20:209–225.

Riedl, R. 1961. Hemichordata. *In:* Handbuch der Biologie. (Bertalanffy and Gessner, eds.) Pt. 6, pp. 409–438. Akademische Verlagsgesellschaft Athenaion.

Tarlo, L. B. 1960. The invertebrate origins of the vertebrates. 21st Int. Geol. Congr. Rept., 21st Session, Norden, Pt. 22:113–123.

Welsch, U., and V. Storch. 1971. Fine structural and enzymehistochemical observations on the notochord of *Ichthyophis glutinosus* and *Ichthyophis kohtauensis* (Gymnophiona, Amphibia). Z. Zellforsch. 117:443–450.

Wilhelmi, R. W. 1942. The application of the precipitin technique to theories concerning the origin of vertebrates. Biol. Bull. 82:179–189.

Williams, J. B. 1960. Mouth and blastopore. Nature 187:1132.

FISHES

Allis, E. P., Jr. 1897. The cranial muscles and cranial and first spinal nerves in *Amia calva*. J. Morphol. 12:487–808.

Allis, J. P., Jr. 1922. Cranial anatomy of *Polypterus*. J. Anat. 56:189–294.

Bigelow, H. B., W. C. Schroeder, and I. P. Farfante. 1948. Fishes of the western North Atlantic. Pt. I.: Lancelets, cyclostomes, and sharks. New Haven, Sears Foundation for Marine Research, Yale University.

Brodal, A., and R. Fange, eds. 1963. The biology of *Myxine*. Oslo, Universitetsforlaget.

Brown, M. E., ed. 1957. The physiology of fishes. 2 vols. New York, Academic Press.

Burger, J. W. 1962. Further studies on the function the rectal gland in the spiny dogfish. Science 131:670–671.

Burger, J. W., and W. N. Hess. 1960. Function of rectal gland in the spiny dogfish. Science 131: 670–671.

Cahn, P. H., ed. 1967. Lateral line detectors. Bloomington, Indiana University Press.

Daniel, J. F. 1934. The elasmobranch fishes. 3rd ed. Berkeley, University of California Press.

Dean, B., 1895. Fishes, living and fossil. New York, The Macmillan Company.

Denison, R. H. 1951. The exoskeleton of early Osteostraci. Fieldiana, Geology 11:197–218.

Denison, R. H. 1956. A review of the habitat of the earliest vertebrates. Fieldiana, Geology 11: 359–457.

Denison, R. H. 1963. The early history of the vertebrate calcified skeleton. Clin. Orthopaedics 31:141–152.

Gans, C., and T. S. Parsons. 1964. A photographic atlas of shark anatomy. New York, Academic Press.

Gregory, W. K. 1933. Fish skulls: a study of the evolution of natural mechanisms. Trans. Amer. Phil. Soc. 23:75–481.

Gregory, W. K., and H. C. Raven. 1944. Studies on the origin and early evolution of paired fins and limbs. Ann. N.Y. Acad. Sci. 42:273–360.

Gunther, Albert C. L. G. 1880. An introduction to the study of fishes. Edinburgh, Adam and Charles Black.

Hardisty, M. W., and I. C. Potter, eds. 1971 and 1972. The biology of lampreys. Vols. I and II. New York, Academic Press.

Hertwig, O. 1874. Über Bau und Entwicklung der Palcoidschuppen und der Zähne der Selachier. Jena Zeitschr. Naturwiss. 8.

Jensen, D. 1966. The hagfish. Sci. Amer. 214:82–90.

Johansen, K. 1968. Air-breathing fishes. Sci. Amer. 219:102–111.

Jollie, M. 1971. A theory concerning the early evolution of the visceral arches. Acta Zool. 52:85–96.

Jordan, D. S., and B. W. Evermann. 1900. The fishes of North and Middle America. Pt. IV. Bull. U.S. Nat. Mus. 47(4).

Marinelli, W., and A. Strenger. 1954 and 1959. Vergleichende Anatomie und Morphologie der Wirbeltiere. I. Lieferung. *Lamperta fluviatilis* L. III. Lieferung. *Squalus acanthias* L. Vienna, Franz Deuticke Verlag.

Millot, J. 1955. The coelacanth. Sci. Amer. 193:34–39.

Moss, M. L. 1964. The phylogeny of mineralized tissues. Int. Rev. Gen. and Exp. Zool. 1:297–331.

Moy-Thomas, J. A. 1971. Paleozoic fishes. 2nd ed., revised by R. S. Miles. Philadelphia, W. B. Saunders Company.

Norman, J. R. 1963. A history of fishes. 3rd ed. London, Ernst Benn, Ltd.

Norris, H. W., and S. P. Hughes. 1920. The cranial, occipital, and anterior spinal nerves of the dogfish, *Squalus acanthias*. J. Comp. Neurol. 31:293–402.

Orvig, T., ed. 1968. Current problems of lower vertebrate phylogeny. Proc. 4th Nobel Symp. New York, Intersciences.

Romer, A. S. 1946. The early evolution of fishes. Quart. Rev. Biol. 21:33–69.

Romer, A. S. 1955. Fish origins—fresh or salt water? Deep Sea Res. (Suppl. 3):261–280.

Schaeffer, B. 1965. The role of experimentation in the origin of the higher levels of organization. Syst. Zool. 14:318–336.

Smith, H. W. 1932. Water regulation and its evolution in fishes. Quart. Rev. Biol. 7:1–26.

Smith, J. L. B. 1939. A living coelacanthid fish from South Africa. Trans. Roy. Soc. South Africa 28(1):1–106.

Thompson, K. S. 1969. The biology of the lobe-finned fishes. Biol. Rev. 44:91–154.

Thompson, K. S. 1971. The adaptation and evolution of early fishes. Quart. Rev. Biol. 46:139–166.

Westoll, T. S. 1958. The lateral fin-fold theory and the pectoral fins of ostracoderms and early fishes. *In:* Studies on fossil vertebrates. (Westoll, T. S., ed.) London, University of London Press.

TETRAPODS

Axelrod, D. I., and H. P. Bailey. 1968. Cretaceous dinosaur extinction. Evolution 22:595–611.

Barcley, O. C. 1946. The mechanics of amphibian locomotion. J. Exp. Biol. 23:177–203.

Barry, A. 1951. The aortic arch derivatives in the human adult. Anat. Rec. 111:221–238.

Bellairs, A. d'A., and G. Underwood. 1951. The origin of snakes. Biol. Rev. 26:193–237.

Brodkorb, P. 1971. Origin and evolution of birds. *In:* Avian biology. (Farner, D. S., and J. R. King, eds.) Vol. I., pp. 19–55. New York, Academic Press.

Carrol, R. L. 1969. Problems of the origin of reptiles. Biol. Rev. 44:393–432.

Colbert, E. H. 1961. Dinosaurs—their discovery and their world. New York, E. P. Dutton and Co.

Colbert, E. H. 1966. The age of reptiles. New York, W. W. Norton & Company.

DeBeer, G. 1954. *Archaeopteryx lithographica.* London, British Museum (Natural History).

Feduccia, A. 1973. Dinosaurs as reptiles. Evolution 27:166–169.

Field, H. E., and M. E. Taylor. 1954. An atlas of cat anatomy. Chicago, University of Chicago Press.

Francis, E. B. 1943. The anatomy of the salamander. London, Oxford University Press.

Gans, C., ed. 1969 and 1970. Biology of the reptilia. 3 vols. New York, Academic Press.

George, J. C., and A. J. Berger. 1966. Avian myology. New York, Academic Press.

Gilbert, S. G. 1968. Pictorial anatomy of the cat. Seattle, University of Washington Press.

Goin, J. C., and O. B. Goin. 1962. Introduction to herpetology. San Francisco, W. H. Freeman & Company.

Gross, C. M., ed. 1966. Gray's anatomy of the human body. 28th ed. Philadelphia, Lea & Febiger.

Heilmann, G. 1926. The origin of birds. New York, D. Appleton-Century Company.

Jayne, H. 1898. Mammalian anatomy. Pt. I.: The skeleton of the cat. Philadelphia, J. B. Lippincott Company.

Johansen, K., and D. Hanson. 1968. Functional anatomy of the hearts of lungfishes and amphibians. Amer. Zool. 8:191–210.

Jollie, M. 1957. The head skeleton of the chicken and remarks on the anatomy of this region in other birds. J. Morphol. 100:389–436.

Jollie, M. 1960. The head skeleton of the lizard. Acta Zool. 41:1–64.

Miller, M. E., G. C. Christensen, and H. E. Evans. 1964. Anatomy of the dog. Philadelphia, W. B. Saunders Company.

Miller, W. S. 1900. The vascular system of Necturus maculosus. Univ. of Washington Sci. Series 2: 211–226.

Moore, J. A., ed. 1964. Physiology of the Amphibia. 3 vols. New York, Academic Press.

Noble, G. K. 1954. The biology of the Amphibia. New York, Dover Publications.

Olson, E. C. 1959. The evolution of mammalian characters. Evolution 13:344–353.

Packard, G. C., and D. L. Kilgore, Jr. 1968. Internal fertilization: adaptive value to early amniotes. J. Theoret. Biol. 20:245–248.

Ranson, S. W. 1959. The anatomy of the nervous system. 10th ed., revised by S. L. Clark. Philadelphia, W. B. Saunders Company.

Reed, C. A. 1960. Polyphyletic or monophyletic ancestry of mammals, or: what is a class? Evolution 14:314–322.

Reighard, J. E., and H. S. Jennings. 1935. Anatomy of the cat. 3rd ed., revised by R. Elliott. New York, Henry Holt and Company.

Romer, A. S. 1944. The development of tetrapod limb musculature—the shoulder region of Lacerta. J. Morphol. 74:1–41.

Romer, A. S. 1956. The osteology of the reptiles. Chicago, University of Chicago Press.

Schaeffer, B. 1941. The morphological and functional evolution of the tarsus in amphibians and reptiles. Bull. Amer. Mus. Nat. Hist. 78: 395–472.

Schmalhausen, I. I. 1968. The origin of terrestrial vertebrates. (Translated from the Russian by L. Kelso; K. S. Thompson, ed.) New York, Academic Press.

Simpson, G. G. 1959. Mesozoic mammals and the polyphyletic origin of mammals. Evolution 13:405–414.

Simpson, G. G. 1960. Diagnosis of the classes Reptilia and Mammalia. Evolution 14:388–392.

Sisson, S. 1953. The anatomy of the domestic animals. 4th ed., revised by J. D. Grossman. Philadelphia, W. B. Saunders Company.

Szarski, H. 1962. Origin of the Amphibia. Quart. Rev. Biol. 37:189–241.

Van Valen, L. 1960. Therapsids as mammals. Evolution 14:304–313.

Watson, D. M. S. 1917. The evolution of the tetrapod shoulder girdle and fore-limb. J. Anat. 52: 1–63.

Watson, D. M. S. 1951. Paleontology and modern biology. New Haven, Yale University Press.

White, F. N. 1968. Functional anatomy of the heart of reptiles. Amer. Zool. 8:211–219.

Wilder, H. H. 1903. The skeletal system of Necturus maculosus Rafinesque. Mem. Boston Soc. Nat. Hist. 5:357–439.

Williams, E. E. 1959. Gadow's arcualia and the development of tetrapod vertebrae. Quart. Rev. Biol. 34:1–32.

Williston, S. W. 1925. The osteology of the reptiles. Cambridge, Harvard University Press. (Reprinted 1971.)

Young, J. Z. 1957. The life of mammals. New York, Oxford University Press.

Index